HUMAN AND ECONOMIC GEOGRAPHY

Dhurata Hajdini

TAP

Toronto Academic Press

HUMAN AND ECONOMIC GEOGRAPHY

Dhurata Hajdini

Toronto Academic Press
4164 Lakeshore Road
Burlington ON L7L 1A4
Canada
www.tap-books.com

Email: orders@arclereducation.com

ISBN: 978-1-77956-296-8

ABOUT THE AUTHOR

Dhurata Hajdini is a graduate of the University of Prishtina and holds an MA in Banking & Finance. Inspired by her economist father since childhood to pursue economics, she now has a particular interest focusing in international economics. When she's not working, she can usually be found walking in nature or reading.

Table of Contents

List of Figures

List of Tables

PREFACE

Humans are always changing the world to fit their needs and desires. Human geography is a study that looks at our languages, cultures, technology, and how we interact with our environment. It's important because it helps us understand different cultures and our own lives better. It tells us about the good and bad things in our environment and how our culture affects us.

Economic geography is a part of geography that studies where economic activities happen, like where businesses are located, and how information, materials, and people move around. It started in the late 1800s but didn't focus on theory much at first, unlike economics. It was more about looking at the real world and seeing how a place's natural and human resources affected its economy. This changed in the 1950s when economic geographers started using data to find patterns in how businesses decide where to be. They looked at things like where markets are, transportation costs, and resources. This helped in understanding why businesses and people make the decisions they do about where to go. This way of studying is also connected to something called 'regional science,' which helps in government planning and solving problems.

The goal of the book "Human & Economic Geography" is to introduce readers to the fascinating world where human activities and economic processes meet the physical landscape. This book is aimed at students and anyone new to the subject who wants to understand how people's lives and economic activities are shaped by and, in turn, shape the world's diverse geographies.

The book breaks down complex concepts into easy-to-understand language, covering topics like population distribution, cultural landscapes, natural resources, and the impact of economic activities on the environment. It aims to show how geography influences economic decisions, such as where to build factories, how trade routes are established, and why cities develop where they do. As it includes real-world examples and illustrations, the book makes learning about human and economic geography engaging and relatable. It encourages readers to think about the interconnectedness of the world and the role geography plays in everyday economic and social decisions. This book is perfect for anyone curious about why the world works the way it does, from why certain areas are more populated to how geography can drive economic growth.

–Author

CHAPTER 1

Basics of Economic Geography

LEARNING OBJECTIVES

After studying this chapter, you will be able to:

- Understand the definition, development and scope of economic geography
- Explain the type of economic activities
- Focus on agglomeration economies
- Discuss about neo-liberalism
- Explain the concept of the digital divide

"Geography has made us neighbors. History has made us friends. Economics has made us partners, and necessity has made us allies. Those whom God has so joined together, let no man put asunder."

-John F. Kennedy

INTRODUCTORY EXAMPLE

Globalization and Economic Geography

The current iteration of globalization can be interpreted as a growing degree of interconnectedness between individuals and locations on both an economic and sociocultural level, fundamentally altering how national economies function. Using the phrase "space-time compression," coined by geographer David Harvey, is one method to comprehend how. The concept of "space-time compression" describes how, as the amount of time it takes to travel through space (both physically and virtually) is decreasing, the world appears to be getting smaller. Is the world becoming a single place as space-time contracts? In certain ways, yes. What happens to the distinctiveness of place as the world gets "smaller"? Some locations share similarities in appearance and texture even though they are very far apart. It is quite possible to come across Americanized fast food with its recognizable packaging while traveling the world. You may hear the same songs that are popular in your hometown. This, however, is far from the whole story; globalization is intensifying place differentiation in many respects. When the spread of culture slows down, regional and ethnic distinctions may become increasingly valuable. The greater their global reach, the more adaptable transnational corporations can be to local variations.

Transnational Corporations (TNCs), which conduct business across borders, and the ensuing labor outsourcing to regions with laxer regulations and lower labor costs are among the factors that have contributed to the current state of globalization. An example of this would be the way that Apple uses labor from maquiladoras in China to make the iPhone, as demonstrated by the Foxconn factory in Zhengzhou, China, or the foreign-owned factories in Mexico that make tax-free goods that are exported from the US. Please refer to the movie synopsis of Maquilapolis for insights into what it's like to work in the maquiladora sector, which primarily employs women. Some view globalization as a form of neocolonialism, whereby countries that were once colonies continue to maintain political and economic dominance over newly independent nations. Unlike the formal, direct rule of core countries over colonies, this rule is informal. Covert intelligence operations, commercial relations, and international financial regulations are how informal rules are enforced. Corporations in core nations expanded and became transnational at the same time, giving them the ability to put financial pressure on PE. Economic geographers frequently use commodity chains, which display the origins of all the constituent parts of a manufactured good, to illustrate the genuinely global reach of modern globalization and international economic integration. Many miles, if not hundreds of thousands, may separate everyday products from their final destination. The results of consumption make this possible. Rich individuals and areas can consume more, while businesses can lower costs through outsourcing. Not only are goods and parts outsourced, but labor and environmental effects as well. Consider cotton production. The United States leads the world in this regard by a wide margin. The majority of clothing made of cotton that was sold in the US, however, was imported. Why would companies that require the raw material to make finished clothing ship it overseas, while the US produces the raw material (with government subsidies)? Through global outsourcing, the company reduces labor costs, and domestically produced clothing reduces greenhouse gas emissions and other pollution in the United States. Following that, the finished product is either returned to the US at no cost to the final customer or shipped back to the US.

INTRODUCTION

With so many different topics to study, geography has developed into a very flexible and diverse subject. As mentioned, assessing the economic achievements of man in terms of production and consumption in relation to his environment is the main objective of economic geography. To assess the relative importance of this area of geography, we must consider the purposes for which it is useful. Place names, natural settings, and how those environments affect human activity have all become outmoded and unwelcome ideas in the context of broader and more modern conceptions of geography.

Geographical differences on the surface of the earth are the subject of modern geography, which is an area-based or spatial science that examines the relationships between various geographic variables. Out of all the major branches of geography, economic geography is the most advanced. It deals with regional variations in human economic activity on the surface of the earth and, despite being a branch of human geography, it now holds a distinct position in geographical studies. Within the field of geography, economic geography has experienced notable transformations over the past half-century, giving rise to a multitude of subfields, including transportation, industrial, and agricultural geography.

1.1. DEFINITION, DEVELOPMENT AND SCOPE OF ECONOMIC GEOGRAPHY

Learning Objectives

- Define economic geography
- Explain the importance of economic geography
- Identify theories of economic geography
- Discuss about world economy
- Understand economic geography examples

Money doesn't grow on trees, but you can make some cash by cutting down trees and selling the wood! Economics has always been a crucial part of human civilization, and economic geography aims to explore how wealth has impacted different regions around the world.

There are various sectors that form an economy, and many theories have been suggested to explain how economies develop.

1.1.1. Definition of Economic Geography

The main goal of economic geography is to identify patterns in the economy over time and space, shedding light on the processes and reasons behind the evolution of economic systems and practices. Economic geography studies the spatial distribution of economic development and activity.

Economic geography is a significant part of human geography as it involves many of the economic transactions that happen in our daily lives. Studying economic geography helps answer questions like: How does economic wealth influence human well-being? What effects does economic development have on human welfare?

Approaching economic geography can be done in numerous ways. The methods are endless: you can look at historical trends, real estate, gentrification, *urbanization*, different economic systems, wealth distribution within nations, regions, and cities, trading philosophies, and policies. However, some economic geography theories have become key pillars of current research. Let's delve into those below.

Keyword

Urbanization is the process through which cities grow, and higher and higher percentages of the population come to live in the city.

1.1.2. Importance of Economic Geography

Because it aids in our understanding of the outside world, economic geography is significant. We can discover more about the variables influencing corporate location decisions by examining the distribution of the economy. We can also learn more about the ways in which various parts of the world interact economically. Furthermore, economic geography can aid in our decision-making regarding trade and transportation. We can make sure that goods and services can move freely between different areas and optimize our transportation networks by understanding how people travel and trade.

1.1.3. Theories of Economic Geography

Understanding the distribution of wealth is a complex topic, and geographers have produced a variety of theories and hypotheses to try to explain broad trends.

Remember

Economic geography is a complex and fascinating field of study. The study of economic geography can help us understand how people and businesses interact with their environment.

1.1.3.1. Weber's Least Cost Theory

The German economist Alfred Weber proposed that industries would probably choose to locate manufacturing plants where labor and transportation would be most affordable, in an effort to minimize costs and maximize profits. The Least Cost Theory is the name given to this pattern. Several Western industries have moved their manufacturing to developing countries where labor costs are lower, thereby implementing the Least Cost Theory.

1.1.3.2. Commodity Dependence and Dependency Theory

A nation that becomes economically dependent on the export of commodities in order to create wealth is said to be in a commodity-dependent situation. It shares some similarities with dependency theory, which is a framework that supports or facilitates commodity dependence—that is, the movement of resources from developing to developed nations. Richer nations can profit from dependency theory and commodity dependence since they give them access to less expensive resources.

1.1.3.3. World-Systems Theory

Did you know?

Brexit is the coined term (a portmanteau) for the exit of the United Kingdom from the European Union.

The core and the periphery were just discussed. These ideas are related to Immanuel Wallerstein's world-systems theory of economics. According to the world-systems theory, countries can be classified into three economic regions: the core, the semi-periphery, and the periphery, depending on which one has greater trading power. Most resources flow from the periphery to the core, where economic wealth is primarily concentrated. The semi-periphery is situated in the middle. The majority of economists and geographers today generally associate the West with the core (Figure 1.1).

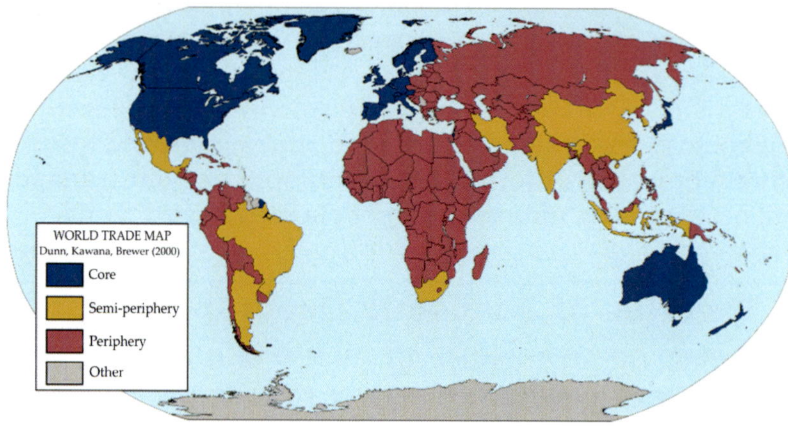

Figure 1.1. The distribution of core, semi-periphery, and periphery countries in world trade as seen at the turn of the century.

Source: By Vladusty – Own work, CC0, https://commons.wikimedia.org/w/index.php?curid=141896621.

1.1.3.4. Rostow's Stages of Growth

In 1960, Walt Whitman Rostow, an American economist (not to be confused with the well-known poet), put forth five stages of economic development in an effort to pinpoint trends in economic expansion.

They are:

1. Traditional society
2. Preconditions for take-off
3. Economic take-off
4. Drive to economic maturity
5. Age of high mass consumption

An eventual catalyst for a country to transition from bartering, hunting, and gathering to sedentary agriculture, industrialization, and ultimately mass consumer spending would arise from social, political, or military demands. The stages identified by Rostow are more appropriate for Western models of economic development. Nevertheless, some may consider it unrealistic that the stages follow a strictly linear progression.

1.1.4. World Economy

It's evident that economic development isn't just a passing trend. It's a key aspect of economic geography. Whether it's through trade, outsourcing labor, or economic and military competition, a country's development is closely tied to its relationships with others, be it cooperation or rivalry.

The world economy, which refers to the interactions of all national economies, is influenced by these dynamics. Neoliberalism, a school of thought promoting international trade, is widely embraced, fostering global interconnectedness. This is often achieved by promoting free trade, eliminating tariffs on imports and exports.

Remember

The world economy includes all of the different interactions between different national economies.

1.1.5. Economic Geography Examples

Let's examine a few distinct development concept examples to see economic geography in action.

1.1.5.1. GDP and the Human Development Index

A nation's level of economic development can be ascertained in several ways. The Human Development Index (HDI), gross national income (GNI), and gross domestic product (GDP) are arguably the three most popular measures.

GDP is the total monetary worth of all economic activity that takes place inside a nation's boundaries. The GDP is divided by the total population residing within the borders to get GDP per capita. Typically, GDP is calculated once a year. Accordingly, the GDP per capita in the United States in 2019 was slightly over $65,000, with a GDP of $21.43 trillion. GNI and GDP are very similar, with the exception that GNI solely tracks the economic activity of citizens and official residents both domestically and internationally. For instance, a citizen of the United States residing and employed in France would be included in the GNI rather than the GDP.

A country's level of economic development is generally indicated by its GDP/GNI ratio. However, this is not always the case; although they are all regarded as developing or semi-peripheral countries, China, India, and Brazil all have GDPs in the top ten.

The Human Development Index, a method of gauging development that combines per capita income with other general quality of life indicators like education and life expectancy, was established by the UN in response to the issue that GDP and GNI do not provide a full picture.

Only one nation, Germany, has both a top 10 GDP and a top 10 HDI ranking. Economic development by itself does not necessarily result in an improvement in the standard of living for the average citizen, as indicated by the difference between GDP and HDI ranking. The idea of *sustainable development* is partly a result of this awareness.

Keyword

Sustainable development is the development that meets the needs of the present without compromising the ability of future generations to meet their own needs.

1.1.5.2. Sustainable Development and Resistance to Modernization

We frequently take the positive effects of economic development for granted. And in most cases, it benefits a large number of people. However, there are drawbacks to economic development as well. Even as the country's wealth increases, uneven development can result in some people living in poverty. Furthermore, as the natural world is exploited for resources, economic development often has a negative impact on the environment.

Enter sustainable development, an idea that aims to strike a balance between respecting human rights, the environment, and long-term viability while also promoting economic growth. To aid countries in achieving global sustainability, the UN has established 17 Sustainable Development Goals.

It is important to note, however, that not all groups adhere to the basic assumptions of economic growth. Some factions, like the Maasai people of Kenya, reject the concept of development entirely and prefer to be left in peace to continue their traditional ways of hunting and gathering or herding cattle on the move, largely disregarding the world economy. Their position contradicts the increasing interdependence of the global economy.

Question

What is the main objective of economic geography?

Answer

The main objective of economic geography is to identify economic patterns across space and time in order to provide insight into how and why economic systems and practices develop.

1.1.6. Development of Economic Geography

Economic geography studies how economic activity is distributed geographically as well as the factors and mechanisms that affect it. Over the past 50 years, economic geography has moved from description—acquiring data about production in different parts of the world—to interpretation, moving from environmental determinism to economic determinism.

Because of its mechanical assumptions about Economic Man and Optimal Location, Neo-classical Economics has been incorporated into economic geography, leading to the development of important fields of study like Industrial Location Theory and Regional Science. On the other hand, a behavioral approach that places more of a focus on the decision-maker has recently been developed.

During the last three decades, economic geography has shifted its focus to:

- The nature and causes of underdevelopment and development, with a focus on the mode of production and an emphasis on the connections between the less developed and more developed worlds.
- The link between economic systems and geography, particularly in interpretations of capitalism's spatial impacts and role in the development of the world economy;
- The impact of technological advancements and the development of new industrial areas
- The use of discursive, qualitative, and realist explanations that acknowledge that each economic agglomeration and development is entrenched locally in its own socio-institutional environment;
- The economic dimensions of class, race, and gender, emphasizing and at times criticizing how economic institutions rely on discrimination based on these three groups;
- The role of 'non-economic' forces in the economic process, such as cultures, institutions, and social behaviors;
- New explanations for the location of economic activity have been proposed, based on economic man's assumptions and optimization;
- The introduction of the concept of the decision-maker and the satisfier in the behavioral approach has changed the ideas; and
- The concept of sustainable development in resource use, etc.

A philosophy of economic geography as well as regional planning and development have resulted from all of these academic disciplines.

Did you get it?

1. **Why is economic geography important?**
2. **Describe the key economic-geographical concepts of space, place, and scale and discuss their usefulness for analyzing contemporary societies and economies.**

1.2. TYPE OF ECONOMIC ACTIVITIES

Learning Objectives

- Understand primary economic activities
- Identify secondary activities
- Explain tertiary, quaternary and quinary activities

The term "economic activities" is used by economists to describe human endeavors that generate revenue. Primary, secondary, and tertiary activities are the three main categories into which economic activities can be separated. Higher services are categorized as tertiary activities and are further separated into quinary and quaternary activities.

If we start by understanding the differences between industries in different economic sectors, it will become easier to grasp the distinctions among industries in other sectors of the economy. This will help us comprehend the factors that influence the distribution of primary, secondary, and tertiary sector industries worldwide.

1.2.1. Primary Sector

Workers in the economy's primary sector extract or harvest goods from the earth, such as basic foods and raw materials. Along with other related activities, some of the activities associated with primary economic activity include mining, forestry, grazing, fishing, hunting and gathering, and agriculture (both commercial and subsistence). This industry is also thought to include the processing and packaging of raw materials.

A declining percentage of workers in both developed and developing nations are employed in the primary sector. For instance, merely 11.8% of the US labor force was engaged in primary sector employment as of 2018. In 1880, about half of the population was employed in the mining and agricultural sectors. This is a significant decline from that number. Primary activity workers are known as "red-collar workers" because they spend the majority of their workdays outside.

1.2.2. Secondary Sector

Using the raw materials extracted by the primary sector, the secondary sector produces finished goods. This sector includes all manufacturing, processing, and construction jobs, with employees often referred to as blue-collar workers. It encompasses various industries such as construction, brewing, automotive and textile manufacturing, chemical and engineering sectors, aerospace manufacturing, energy utilities, and shipbuilding. In 2018, 12.7% of the US working population was involved in secondary sector activities.

Remember

Production, consumption, and capital formation are the primary economic activities of an economy.

1.2.3. Tertiary Sector

The "service industry" or "service sector" is another name for this economic sector. The third sector sells goods and services generated by the secondary sector to businesses in all five economic sectors as well as to the general public.

Numerous industries are included in this sector, including retail and wholesale sales, distribution and transportation, dining establishments, secretarial services, media, *tourism*, insurance, banking, healthcare, and legal practice.

The percentage of workers in both developed and developing nations who are employed in the tertiary sector is rising. Approximately 61.9% of all workers in the United States are in the tertiary sector. The Bureau of Labor Statistics divides the non-agricultural self-employed into a different category, which includes 51.6% of workers; however, the industry in which they operate depends on their job.

Keyword

Tourism is a fascinating blend of social, cultural, and economic aspects, encompassing the travel of individuals to different countries or locations beyond their usual surroundings for personal or business/professional purposes.

1.2.4. Quaternary Sector

Because they are so specialized, quaternary activities in the "Knowledge Sector" need to be classified separately from tertiary activities. Information-based services are in much greater demand and are consumed by a wider range of people these days, including statisticians, software developers, mutual fund managers, and tax advisors. This group of services includes workers in office buildings, classrooms, and elementary schools, theaters, medical offices and hospitals, accounting and brokerage firms, and other establishments of a similar nature. Quaternary tasks can be contracted out similarly as some tertiary functions. They don't depend on resources, are unaffected by the environment, and aren't always limited to a certain region.

1.2.5. Quinary Sector

Quinary activities are those that involve the development, reorganization, and interpretation of both new and old concepts;

interpreting data, and utilizing and assessing emerging technologies. Because they represent a different segment of the tertiary sector that embodies the specialized and highly compensated skills of, among other people, senior business executives, government officials, research scientists, financial and legal consultants, and other professionals, they are sometimes referred to as "gold collar" jobs. The fact that they are few in number does not diminish their significance in the architecture of advanced economies. Decision-makers or policymakers with the highest authority carry out quinary activities.

Some economists further subdivide the quaternary sector into the quinary sector, which comprises the highest levels of decision-making in a society or economy. Prominent figures from various industries, including government, science, higher education, nonprofits, healthcare, culture, and the media, are employed in this field. Since police and fire departments are regarded as public services rather than for-profit businesses, they might also fall under this definition.

At home, activities such as household tasks carried out by a family member or dependent are occasionally categorized as belonging to the quinary sector. These activities, which include housekeeping and child care, have an economic impact on the economy, even though their value cannot be measured in monetary terms because they provide services that would otherwise be charged for. A little over 13% of all workers in the US are employed in the quinary sector. Those in quinary occupations are referred to as gold collars (Figure 1.2).

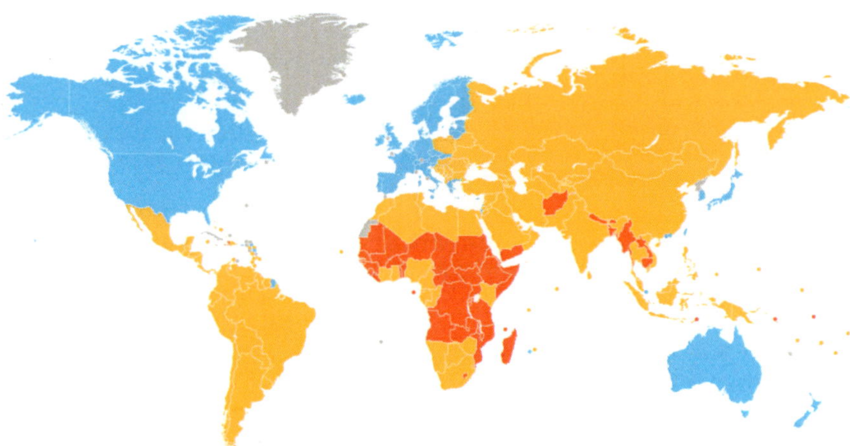

Figure 1.2. There is often a relationship between economic sectors and the level of economic development [*Note:* Developed countries are in blue; developing countries are in yellow; and the least developed countries are in orange].

Source: By Allice Hunter - Blank map: File: World map (Miller cylindrical projection, blank). svgIMF data: World Economic Outlook Database April 2023 — WEO Groups and Aggregates Information. www.imf.org. Retrieved on 2022-06-02.UN list: UN List of Least Developed Countries as of December 2023. www.un.org., CC BY-SA 4.0, https://commons.wikimedia.org/w/index.php?curid=118625873.

The term "economic activities" refers to revenue-producing human endeavors. Economists categorize economic activities into primary, secondary, and tertiary activities. Tertiary activities include higher services, which are divided into quinary and quaternary activities. Raw materials extracted by the primary sector are used to produce finished goods in the secondary sector. The tertiary sector, also known as the "service industry" or "service sector," focuses on services. Quaternary activities are specialized within the "Knowledge Sector," requiring a distinct classification due to their extreme specialization. Quinary activities involve the development, reorganization, and interpretation of concepts, data, and technologies.

Question

How are the economic activities of a country measured?

Answer

Nominal or real Gross Domestic Product (GDP) indicators are used to quantify a nation's economic activity. The GDP represents the total value of goods and services produced by an economy within its boundaries during a specific time period. Factors like technology, infrastructure, natural resources, and the labor force all influence a nation's economic activity. Developed countries tend to focus on tertiary activities, while developing nations often prioritize primary industries. Tertiary or service industries contribute significantly to the global GDP, accounting for over half of it.

A nation will naturally begin to progress as its tertiary and quaternary activity levels increase. This usually happens after industrialization steps that boost a nation's development and help it move towards higher economic activity levels. Progress tends to be much slower when relying solely on primary and secondary activities.

Example

Let's take a look at the economic activities of the UK and Bangladesh. The UK shifted quickly from a secondary activity-based economy to mainly tertiary activity due to its early industrialization. This allowed the country enough time to develop into a tertiary and quaternary-dominated economy, helping the British allocate resources effectively. On the other hand, Bangladesh heavily depends on exporting primary and secondary products like rice and clothes. With limited capital, it faces challenges in speeding up development. As a result, Bangladeshi citizens generally have lower disposable incomes compared to UK citizens. Additionally, Bangladesh's resources are mainly focused on primary and secondary industries, leaving little for domestic growth, leading to slow economic progress.

Did you get it?

1. **What are some examples of economic activities?**
2. **What is economic activity and its types?**

1.3. AGGLOMERATION ECONOMIES

Learning Objectives

- Understand the types of economies
- Explain core-periphery model
- Identify the source of economies
- Focus on the advantages of agglomeration
- Know the disadvantages of agglomeration

Economic activity tends to cluster in specific areas known as agglomerations. This is one of the most extensively studied aspects of the spatial economy's structure, observed across various geographic scales worldwide and throughout history. The growth of cities, the development of industrial districts and regions, and the concentration of related activities in a town or city's vicinity are all examples of this phenomenon. This is why national economies often concentrate in specific areas rather than being evenly spread across their territories. This section delves into the benefits of agglomeration, its varying impacts on different regions' economies, and its effects on productivity.

1.3.1. Types of Economies

Localization and urbanization economies are two examples of large-scale economies with external economies of scale. Localization economies are the result of numerous businesses in the same industry setting up shop nearby. Localization economies come from three sources. The first is the advantage of labor pooling, which is the availability of a range of skilled workers to businesses, which in turn gives workers job opportunities. The second advantage stems from the growth of industries brought about by the increasing returns to scale in intermediate inputs for a product. Meanwhile, the third advantage stems from the comparatively easy communication and exchange of labor, supplies, and innovative ideas resulting from the close proximity of firms.

1.3.2. Core-periphery Model

Although the localization and urbanization economies, as well as their sources, are essential to the survival of agglomeration economies and cities, it is critical to comprehend the long-term outcome of agglomeration economies' function in relation to the core-periphery model. A remote region with less dense activity is encircled by a large amount of economic activity in the core-periphery model. Since businesses in these agglomeration economies choose locations that minimize costs to maintain high productivity and other advantages, the concentration of economic activity in one area—typically a city center—allows for the growth and expansion of activity into other and surrounding areas. As a result, these businesses can expand beyond the city (core) and into the periphery. A slight reduction in the production cost can broaden the area where businesses can locate themselves, potentially causing a loss of concentration within the city and the creation of a new city outside the area where agglomeration and growing returns to scale are present.

Each company in the same industry should form its own city if localization economies account for the majority of the reasons cities exist, independent of urbanization economies. But in reality, cities are more complicated than that, which is why large cities are created by the interaction of the localization and urbanization economies.

Did you know?

The most important drivers of agglomeration are an educated workforce and a skilled labor pool, followed by local input suppliers.

1.3.3. Source of Economies

Labor market pooling results from the localization of businesses. Many skilled workers move into the area, allowing for the exchange of information, ideas, and knowledge. Higher wages for the workers are the outcome of increased competition among the firms in this area to attract workers. Nonetheless, the wage for those workers decreases with the number of businesses and the number of employees at a given location.

The availability of specialized products and services for the clustering firms is the second factor contributing to localization economies. Intermediary inputs are those that provide access to specialized goods and services. Because of its proximity to readily available sources required for production, it offers each company operating there increasing returns to scale. A core-periphery concept with tradable intermediate inputs will have numerous businesses positioned close to one another in order to be nearer to their needed sources. Let's say there are tradeable goods and services in the vicinity, but no associated industries exist in the same region. The lack of networking connections in that scenario makes it challenging for all local businesses to acquire resources and boost output. Because the clustering of firms results in lower transportation costs, there is a greater chance of a core-periphery pattern. This means that more intermediate inputs will be concentrated at the core, which will draw in more firms in related industries.

Spillovers from technology represent the third source of localization economies. Additionally, clustering in specific fields accelerates the spread and adoption of ideas, offering another advantage of this source. Businesses require realistic access to capital markets to operate at full capacity and market their products effectively. However, new technologies can pose risks and challenges. By sharing information, a cluster of businesses can mitigate these challenges, making it easier to adopt new technologies with less uncertainty. Locating close to these areas is beneficial for companies due to the concentration of technology and capital flow industries. This technological impact leads to a higher concentration of information flow and economic activity, particularly in the communications field, breaking down barriers between businesses in the same industry, whether nearby or farther away. Moreover, smaller cities have already reaped the benefits of forming and expanding informational networks, potentially allowing technological spillovers to have a more significant impact on their development compared to larger cities.

1.3.4. Advantages of Agglomeration

Specific development strategies are implemented within and across economic activity clusters formed by firms. To achieve what economists refer to as increasing returns to scale, this facilitates the flow of fresh and creative ideas among firms as well as the accumulation of information. A company's *internal* economies of scale, or increasing returns to scale, can facilitate the expansion of the company into new markets. Agglomeration economies of scale, or economies of scale *external* to a firm, are the outcome of spatial proximity. Agglomeration economies can be understood as the internal and external environment of the region and businesses, respectively. According to Beckmann, a key to comprehending the formation of urban centers is the increasing returns to scale. The trade-off between transportation costs and economies of scale is captured by these growing returns to scale, which "give rise to [urban systems]." When production is less expensive as a result of this concentration of economic activity, agglomeration economies occur. The establishment of additional companies that may benefit from these economies without joining any large organization is made possible by this clustering. Regions may become more urbanized as a result of this process.

Benefits arise from the spatial agglomeration of physical capital, companies, consumers, and workers:

- *Low transport costs:* Physical proximity to other firms and centers of production can minimize costs associated with transportation. While this may have been the case for many manufacturing firms in the United States, Glaeser and Gottlieb argue that reducing transportation costs is more important for firms producing services. Moreover, other studies have shown that when negative externalities like pollution are taken into account, agglomerated city centers are more likely to be dispersed over a larger geographical area rather than be confined to a single, metropolis-like urban region.
- *Geographic advantages*: A longstanding theory in urban economic literature is that firms

(and cities) are likely to agglomerate where there are natural geographic advantages, which would give these firms both comparative and cost advantages over their competitors. Ellison and Glaeser argue that while this may be true for firms whose location decisions are highly sensitive to cost differences or geographic locations, such as the wine industry, they find that only 20% of geographic agglomeration effects in the United States can be explained by "natural" cost advantages.

- *Labor pooling and matching:* Agglomerating effects, such as an increase in population and therefore, human capital, arguably help improve matching within the economy, e.g. employees with employers, suppliers with buyers, and so on. Moreover, massive urban areas like cities, which contain a multitude of industries in a localized area, can help firms offset their reaction to shocks more efficiently by 'pooling' labor resources together.
- *Knowledge spillovers:* The accumulation of knowledge and human capital in concentrated areas such as major urban centers can contribute to the sharing of production technologies (i.e., know-how) between firms. Econometric analysis by Liang and Goetz showed that agglomeration effects account for technology-intensive industries benefiting from Jacobs-type knowledge spillovers. Furthermore, agglomerated centers of production, such as cities, also facilitate learning – that is, knowledge generation, diffusion, and accumulation – on a larger scale than smaller economic regions.

1.3.5. Disadvantages of Agglomeration

Agglomeration could also lead to traffic jams, pollution, and other negative externalities due to the concentration of businesses and people. These unfavorable effects would create economies of scale. Cities can only prosper if their benefits outweigh their drawbacks. In fields or industries where access to relevant production facilities or resources is limited, agglomeration diseconomies can also arise from longer waiting times and overcrowding. As mentioned earlier, the presence of numerous competitors in the market, labor scarcity, and businesses' inflexibility with their workforce are reasons why firms' pricing power diminishes. The balance between agglomeration economies and agglomeration diseconomies can influence the growth of an area, leading to the challenges that large cities encounter. It has also been shown that agglomeration economies worsen inequality within cities and between cities and rural areas. According to the Henry George theorem, Oxford development economist Paul Collier has proposed that taxing the benefits of agglomeration as rents would discourage behavior that distorts markets. Aligning profits with merit and curbing rent-seeking would be both ethical and practical. Collier recommends a tax calculated by aggregating the income of affluent and urban areas, which can then be redistributed to cities greatly affected by agglomeration.

- *For the environment*: Environmental pollution, such as that found in the air, water, and land, is most likely caused by the high degree of economic agglomeration. The economic agglomeration in China's Yangtze River Delta is one example.

In addition to causing some extreme weather problems, the Yangtze River Delta's severe air pollution has increased certain diseases due to the dense population and industrialization. Environmental issues and pressures can be reduced and alleviated by modifying the spatial organization of the city. An additional environmental issue is energy consumption. Because of the economy's agglomeration, industries and population growth grew quickly, posing major energy-related issues. In addition, it would increase the need for production technology and result in some pollution.

- *For the economy*: Agglomeration can have some positive effects on the economy, but it also tends to exacerbate interregional inequality and the gap between rich and poor areas. Because it is a necessary stage of economic development, interregional inequality cannot be avoided. Political and social instability are even brought on by the severe inequality problem. To increase equality, politics must be involved. Examples of this include funding innovation and human resources. Rural areas may also be impacted by economic agglomeration. The excessive density of population in the city would have an impact on agricultural output and lead to issues with unemployment.
- *For society*: Some social problems were also brought about by economic agglomeration. First, the rising population drove up the cost of land. The government may tax the land in the agglomeration area, raising the price of land, in an effort to lessen the inequality issue. Secondly, the economy's agglomeration also raises the demand for infrastructure. The public goods problem must be coordinated, the optimal amount of public contribution must be determined, and inadequate or overburdened infrastructure must be avoided.
- *For companies*: Economic agglomeration would cause fierce competitive pressure. Competition can reduce the market price, motivating companies to innovate and increase production efficiency. However, over-competition would hinder their development and innovation, and also generate some social problems.

Question

What is an example of agglomeration?

Answer

A good illustration of this concept is a shopping mall. Even though the stores in the mall may not be related, being situated close together enables them to share the same infrastructure: the building, parking lots, and other common areas.

Did you get it?

1. **What is agglomeration in regional economics?**
2. **What is the difference between agglomeration and cluster?**

1.4. NEO-LIBERALISM

Learning Objectives

- Understanding neoliberalism
- Explain the history of neoliberalism
- Know the characteristics of neoliberalism
- Describe criticism of neoliberalism

A political and economics-based policy framework is neoliberalism. It supports private industry and aims to give the private sector more control over economic variables than the government does. In order to ensure that free market capitalism operates as efficiently as possible, many neoliberal policies concentrate on reducing public ownership, government regulation, and spending. Margaret Thatcher, the former Prime Minister of the United Kingdom, is frequently linked to neoliberalism. Ronald Reagan, the 40th President of the United States from 1981 to 1989, served as leader of the Conservative Party from 1975 to 1990 and as President from 1979 to 1989.

In contemporary times, neoliberalism has been linked to austerity measures and endeavors to reduce public expenditure on social services.

1.4.1. Understanding Neoliberalism

Globalization, deregulation, free trade, and cutting back on government expenditure are all key components of the political and economic ideology known as neoliberalism. It's associated with laissez-faire economics, a school of thought that advocates for the least amount of government intervention in matters pertaining to the economic well-being of both individuals and society.

According to laissez-faire economics, sustained economic growth will result in increased free market expansion, technological innovation, and minimal government intervention.

Moreover, libertarianism and neoliberalism are occasionally confused. Neoliberals, on the other hand, usually support greater government involvement in the business and social sphere than libertarians do. Neoliberals, for instance, typically support progressive taxation, but *libertarians* frequently reject this position in favor of policies like a flat tax rate for all taxpayers.

Furthermore, libertarians despise neoliberals for frequently supporting policies like major industry bailouts. Though many people's political views are conservative and neoliberal policies run counter to that, it should be highlighted that numerous parties supported neoliberal policies like free trade agreements, the deregulation of the banking, energy, and transportation sectors, and the reduction of the capital gains tax and top income tax rate.

Keyword

Libertarians are classical liberals who strongly emphasize the individual right to liberty.

1.4.2. History of Neoliberalism

In the past 25 years, neoliberalism has become more popular. A new form of liberalism is referred to as neoliberalism. When Scottish economist Adam Smith published his book "The Wealth of Nations" in 1776, the concept of a liberal economy gained popularity.

Smith argued in his book that deregulation and free trade are the two most effective ways to promote economic growth and advocated for the elimination of government intervention.

Many years later, in the 1930s, an additional economist by the name of John Keynes questioned liberalism and concluded that capitalists would benefit from such a course of action. Several leaders changed course and began working for the common good as a result of government intervention, which was seen as essential for the welfare of the populace. But in light of the fall in corporate profitability, the idea of neoliberalism is currently being applied globally.

1.4.3. Characteristics of Neoliberalism

The idea at the heart of neoliberalism is that more economic freedom translates into more personal economic and social advancement. It supports the following claims:

- Free enterprise, competition, deregulation, and the importance of individual responsibility
- Opposition to the expansion of government power, state welfare, inflation
- Minimizing government control of industry and boosting private sector ownership of business and property
- Free market capitalism and the efficient allocation of resources
- Globalization rather than heavily regulated markets and protectionism

- A reduction in government spending and lower taxes
- Less government control over economic activity to enhance the efficient functioning of the economy
- An increase in the impact of the private sector on the economy
- A reduction in union power and greater flexibility in employment
- Government intervention when it's needed to help implement, sustain, and protect free market activities

Did you know?

President Jimmy Carter's deregulation of the airline industry in 1978 is an example of a neoliberal policy in action. The Airline Deregulation Act removed government control over fares, routes, and who could enter the market.

1.4.4. Liberalism vs. Neoliberalism

Liberalism is basically a comprehensive political theory. It covers all aspects of society – social, political, and economic – as well as the role of the government, and it places a high value on liberty. Neoliberalism, on the other hand, is more focused on economics. Neoliberal policies concentrate on markets and the laws and regulations that impact the economy.

1.4.5. Criticism of Neoliberalism

There are several criticisms of neoliberalism outlined below.

1.4.5.1. Misguided Free Market Approach to Public Services

One common critique is that promoting a free market approach in sectors like health and education, which are public services, is misguided. The pursuit of profit in public services is different from that in other industries.

Furthermore, implementing a free market strategy in education and health can exacerbate inequality and lead to underfunding of these crucial resources, which are essential for the long-term health and sustainability of an economy.

1.4.5.2. Monopolies

In the West, neoliberal policies have been implemented at the same time that income and wealth inequality have increased. While highly skilled workers might be able to demand higher pay, low-skilled workers are more likely to experience pay stagnation. Neoliberal policies often contribute to the existence of monopolies, which boost corporate profits at the expense of consumer benefits.

1.4.5.3. Increased Financial Instability

Capital deregulation has not always aided in economic development, despite what neoliberal advocates frequently assert. Rather, the deregulation of capital has resulted in a rise in financial instability, accompanied by occasional global economic shocks. In fact, a report on neoliberalism by the International Monetary Fund (IMF) indicates that the risk of unfavorable economic cycles has increased due to an increase in capital flows.

1.4.5.4. Inequality

It's a proven fact that neoliberal policies make inequality worse. The long-term growth prospects of an economy may be hampered by this inequality. Individuals with low incomes have restricted purchasing power, at one extreme of the spectrum. However, there is a positive correlation between wealth and saving behavior. Contrary to what neoliberal advocates predict, wealth does not trickle down in the latter case.

1.4.5.5. Globalization

Ultimately, globalization, which critics argue takes away the self-determination of sovereign nations, has been supported by neoliberalism's focus on economic efficiency. Critics of neoliberalism argue that it could reduce efficiency by promoting the privatization of state-owned businesses. They suggest that while privatization might increase productivity, the limited geographical area of the world may hinder long-term benefits. Opponents of neoliberalism also argue that it is undemocratic, could criminalize poverty, and lead to exploitation and social injustice.

Question

What Are the Effects of Neoliberalism?

Answer

A few effects could be more open markets, increased access to goods and services to satisfy consumer demand, increased income, and increased profits. Reduced prices as a result of increased competition may also occur. Allocating resources more efficiently can lead to savings. Neoliberal policies can also lead to improved workforce organization and the capacity to hire the right people for the right jobs. Others may draw attention to a few of the detrimental consequences of neoliberalism that are thought to exist. A few of these might be rising monopolies, job insecurity, outsourcing-related job losses, economic inequality, and a growing disregard for the wants and welfare of people.

Did you get it?

1. **What Is Neoliberalism in Simple Terms?**
2. **What is an example of Neoliberalism?**

1.5. CONCEPT OF DIGITAL DIVIDE

Learning Objectives

- Identify types of digital divide
- Explain the causes of the digital divide
- Explain the impact of the digital divide
- Focus on strategies for bridging the digital divide

The difference in access to contemporary information and communication technology between those with and those without is known as the "digital divide." Globally, digital inequality is influenced by three major stages.

Urban and rural communities, socioeconomic groups, less developed and more developed nations, and the educated and uneducated populace all exhibit clear signs of digital inequality. It is possible to digitally divide people who have access to a broadband connection. How? The difference is widened by subpar computers, sluggish internet, and restricted access to content that requires a subscription.

1.5.1. Types of Digital Divide

Our attempts to access the internet are impacted by various forms of the digital divide. The following are a few striking examples of digital inequality:

1.5.1.1. Gender Divide

The gender gap on the internet is quite noticeable, especially in developing nations, as reported in 2013. Despite the rapid growth of mobile connectivity, its distribution remains uneven. Women are still lagging behind. In low-income countries, men are ninety percent more likely than women to own a mobile phone, leaving 184 million women without access to mobile internet. In low- and middle-income countries, 11.2 billion women lack internet access, even among those who own mobile phones.

1.5.1.2. Social Divide

Regarding the social aspect, people with similar interests tend to build relationships and social circles online. Social media platforms like Facebook and Twitter facilitate the formation of online peer groups based on shared interests. Internet usage has played a significant role in shaping *social stratification*, particularly in societies where internet connectivity varies. Non-connected groups are marginalized due to their lack of access to the advantages enjoyed by connected groups.

1.5.1.3. Universal Access Divide

When it comes to using the internet, people with physical disabilities frequently face disadvantages. Despite possessing the required abilities, they are unable to take advantage of the hardware and software that are readily available. Due to a lack of digital literacy, low levels of education, and inadequate broadband infrastructure, some parts of the world will continue to be isolated from the internet and its immense potential.

Keyword

Social stratification refers to a society's categorization of its people into rankings based on factors like wealth, income, education, family background, and power.

1.5.2. Causes of the Digital Divide

Even though more people have access to computers and the internet than ever before, the digital divide is still widening at an alarming rate for the following reasons.

1.5.2.1. Education

In order to close the digital divide, education is a major investment. The gap in digital inequality is getting wider due to low literacy. Compared to those with only a high school education, people with college degrees are thought to be 10 times more likely to utilize computers and the internet to the fullest extent possible in their daily lives.

1.5.2.2. Income Levels

The digital divide is exacerbated in large part by the disparity in income. High-income earners ($75,000) are 20 times more likely to access the internet than low-income earners ($30,000). Compared to low-income families, wealthy families are ten times more likely to own computers and a high-speed internet connection at home. For low-income populations, money is scarce. Their earnings are channeled towards basic needs. They view technology as a luxury.

1.5.2.3. Geographical Restrictions

People in more developed nations have access to a wide range of technologies and high-speed broadband connections. However, in less economically developed nations, the infrastructure and technology needed for high-speed internet are lacking. Additionally, physical barriers within a country can contribute to the digital divide. Urban areas tend to have better access to 4G or fiber optic internet compared to rural or mountainous regions.

1.5.2.4. Motivation and General Interest

Some individuals have the necessary resources and skills but lack interest in computers or the internet, viewing them as unnecessary luxuries. Others find technology too complex to grasp.

1.5.2.5. Digital Literacy

In developed countries, computers and high-speed internet are more readily available. Students who acquire digital skills early on and attend schools with sufficient computer resources have an advantage over those without such exposure.

The information gap between those with access to technology and those without widens, especially in developing nations where physical access to technology is limited.

1.5.3. Impact of the Digital Divide

Globally, there are now noticeable differences in societies due to digital inequality. Some of the most striking effects of the digital divide are as follows.

1.5.3.1. Impact of Digital Divide on the Economy

The economy grows thanks to telecommunications services. An economically productive country is made possible by widespread internet use. People can shop online, doing away with the hassle of rush hour traffic. One practical way to promote economic empowerment is through paperless transactions. Particularly in developing nations without sufficient Information and Communications Technology (ICT) integration, the economic divide grows.

1.5.3.2. Impact of Digital Divide on Education

The internet is a vast informational library. There are many educational platforms available now more than ever for learning advanced skills. ICT access has been connected to superior scientific research and academic success. Knowledge gives the mind power. Keeping up with this rapidly changing industry is essential for both individual and societal success.

1.5.3.3. Impact of Digital Divide on Social Spheres

Communication is driven by internet access. Social media sites like Facebook foster interaction, connections, and keep us in touch with our loved ones. Information can be accessed more quickly than through traditional newspapers.

Technology has had a profoundly revolutionary impact on music and art. Societies without access to the internet do not enjoy these advantages. For developing nations to be empowered, the digital divide must be eliminated.

1.5.3.4. Impact of Digital Divide on Society

The division of people in society based on factors such as gender, age, race, and ethnicity has been exacerbated by the digital divide. Between people who have access to the internet and those who do not, technology produces new alignments. The growth and development of those with restricted access are impeded by their continued lag.

1.5.4. Strategies for Bridging the Digital Divide

One of the Sustainable Development Goals (SDG 9) set forth by the UN is to reduce the digital divide. For this reason, programs to make technology more accessible have been started in many locations. Some of the most pertinent ones are listed here:

- **Digital literacy programs**. They instruct people in less-favored areas on Internet use to improve their personal well-being.
- **Alliance for Affordable Internet (A4AI).** This project, led by an international coalition of governments, businesses, and civil society, aims to lower the cost of broadband in specific areas in Africa, Asia, and Latin America.
- **Free Basics**. This initiative, promoted by Facebook and six other technology companies, aims to provide free access to some websites through a mobile application.
- **Starlink**. This project, promoted by tycoon Elon Musk, is launching satellites into space to provide high-speed Internet and global coverage at affordable prices.

Question

When Did the Term "Digital Divide" Originate?

Answer

The term has been around since the late 20th century when the difference between people with cell phones and those without them was labeled. Today, it refers to the difference between those who have internet access (as well as access to other forms of digital communication) and those who do not.

Did you get it?

1. **What is the theory of the digital divide?**
2. **What are the 3 stages of the digital divide?**

ROLE MODEL

NAOMI KLEIN

Canadian Author and Activist

Naomi Klein (born May 8, 1970, in Montreal, Quebec, Canada) is a Canadian author and activist. Her debut book, *No Logo* (2000), propelled her to become one of the most prominent voices in the anti-globalization movement.

Klein was born into a politically active family. Her grandfather, an animator for Disney, was fired and blacklisted for attempting to organize a labor union. Her parents relocated to Canada from the United States to protest the Vietnam War. Her mother, Bonnie, directed the feminist antipornography documentary *Not a Love Story* (1981). Klein pursued studies in philosophy and literature at the University of Toronto but left before completing her degree to work at the Toronto newspaper *The Globe and Mail*. In 2000, Klein published *No Logo*, an analysis of the marketing and branding practices of global corporations. The book delved into how contemporary capitalism aimed to reshape individuals' consciousness along branded lines. *No Logo* was translated into numerous languages, propelling Klein to international media stardom. She later released *Fences and Windows* (2002), a collection of essays on anti-globalization topics covering events from World Trade Organization protests to an exploration of the Zapatista uprising in Chiapas, Mexico. Collaborating with her husband, director Avi Lewis, Klein co-wrote and co-produced *The Take* (2004), a documentary depicting the occupation of a closed auto-parts plant by Argentine workers. Klein's *The Shock Doctrine* (2007) provided a scathing critique of neoliberalism, particularly targeting Milton Friedman's "Chicago school" of economics. The book examined what Klein labeled "disaster capitalism," an extreme form of capitalism advocating privatization and deregulation following war or natural disasters. *The Shock Doctrine* was adapted into a feature-length documentary film by director Michael Winterbottom in 2009.

In *This Changes Everything* (2014), Klein iterated the inherent conflicts between unchecked capitalist enterprise and the mitigation of global warming; a documentary based on the book and directed by Lewis was released in 2015. *No Is Not Enough: Resisting Trump's Shock Politics and Winning the World We Need* (2017) was written in response to the administration of U.S. Pres. *Donald Trump*. Klein's later books included *The Battle for Paradise: Puerto Rico Takes on the Disaster Capitalists* (2018), which focused on the struggle over the island's recovery following Hurricane Maria. In the essay collection *On Fire: The (Burning) Case for a Green New Deal* (2019), she continued to address the climate crisis.

Klein's work appeared in publications such as *The Globe and Mail*, *The Guardian*, and *The Nation*.

FOCUS ON CAREERS

ECONOMIC ANALYST

Economic analysts research significant trends in commerce and business. From their research, they forecast the effects of these trends on a company or industry. They often work for privately held companies, financial institutions, or state and federal government. Some of the information they might analyze includes energy costs, taxes, healthcare costs, inflation, or employment levels, depending on their industry.

They need to have keen skills in analysis and data interpretation and the ability to identify new trends. Economic analysts should also possess excellent writing and review skills to construct reports that explain their key findings on a topic or industry.

What Economists Do

Economists conduct research, prepare reports, and evaluate issues related to monetary and fiscal policy. They also may collect and analyze statistical data.

Work Environment

Economists typically work in an office setting, either independently or collaborating with a variety of other workers. Most economists work full-time, and some work more than 40 hours per week.

How to Become an Economist

Economists typically need at least a master's degree to enter the occupation. However, some economists—primarily in government—qualify for entry-level positions with a bachelor's degree. Others need a Ph.D.

Pay

The median annual wage for economists was $113,940 in May 2022.

Job Outlook

Employment of economists is projected to grow 6% from 2022 to 2032, faster than the average for all occupations.

About 1,200 openings for economists are projected each year, on average, over the decade. Many of those openings are expected to result from the need to replace workers who transfer to different occupations or exit the labor force, such as to retire.

State & Area Data

Explore resources for employment and wages by state and area for economists.

Similar Occupations

Compare the job duties, education, job growth, and pay of economists with similar occupations.

A CLOSER LOOK

THE RISE OF MORE SOPHISTICATED COMMUNICATIONS TECHNOLOGIES AND HOME WORKING

Before COVID-19 hit, it seemed that more people could work from home thanks to advancements in communications technologies. The percentage of workers who primarily work from home more than tripled between 1980 and 2010, from 0.75% in 1980 to 2.4% in 2010, according to a 2012 US Census Bureau report. A wide range of occupations, from sales assistants to managers and engineers, were performed by home workers, and their pay ranged accordingly.

In the UK, home workers made up more than 13% of the total workforce in 2011, an increase of 2%age points from the previous year. Sixty-three percent of these home workers were self-employed. Although the Netherlands and Finland reported even higher maximum levels of WFH, the WFH rates in the UK were comparable to the EU average. The percentage of home workers increased to 14.5% in the second half of 2019, according to the UK Labor Force Survey (LFS). London's share, at 14.3%, was extremely close to the national average.

Additionally, the LFS discovered that the WFH rates varied systematically by industry and occupation. Professional traders reported the highest WFH rates (25%), followed by managers, directors, and senior officials (21%), while sales and customer service (4.2%) and elementary occupations (2.1%) reported the lowest WFH rates. Cleaners (6%), on the other hand, gave the lowest ratings. This was evident in the qualification of home work. According to a different study, just over 10% of people without any formal education have worked from home, compared to nearly half of the graduates.

Although there has been a slight increase in home working, the concentration of national economies in large cities has not changed. This indicates that jobs in prosperous cities may have seen an increase in hybrid working before the pandemic, or jobs outside of cities were more likely to shift to remote work arrangements.

SUMMARY

- Finding patterns in the economy across time and space is the primary goal of economic geography, as it sheds light on the processes and reasons behind the evolution of economic systems and practices.
- The world economy is the term used to describe the collective interactions of all the various national economies in the world.
- A nation's level of economic development can be determined in various ways. The Human Development Index (HDI), gross national income (GNI), and gross domestic product (GDP) are considered the three most popular measures.
- Economic geography studies how economic activity is distributed geographically as well as the factors and mechanisms that affect it.
- The term "economic activities" is used by economists to describe human endeavors that produce revenue. Primary, secondary, and tertiary activities are the three main categories into which economic activities can be separated. Higher services are categorized as tertiary activities and are further separated into quinary and quaternary activities.
- Workers in the economy's primary sector extract or harvest goods from the earth, such as basic foods and raw materials.
- The secondary sector of the economy produces finished goods using raw materials extracted by the primary sector. This industry encompasses all manufacturing, processing, and construction jobs, with employees often referred to as blue-collar workers called blue-collar workers.
- The service industry, also known as the service sector, sells goods and services from the secondary sector to businesses across all economic sectors and the general public.
- Quaternary activities in the "Knowledge Sector" need to be classified separately from tertiary activities. Information-based services are in much greater demand and are consumed by a wider range of people these days, including statisticians, software developers, mutual fund managers, and tax advisors.
- Quinary activities are those that involve the development, reorganization, and interpretation of both new and old concepts; interpreting data; and utilizing and assessing emerging technologies.
- Economic activity tends to cluster in areas called agglomerations. This is one of the most well-documented aspects of the spatial economy's structure, observable across various geographic scales globally and throughout history. Localization and urbanization economies are examples of large-scale economies with external economies of scale.
- Globalization, deregulation, free trade, and reducing government spending are all important aspects of the political and economic ideology called neoliberalism. It's linked to laissez-faire economics, which promotes minimal government involvement in economic matters affecting individuals and society.

- The difference in access to contemporary information and communication technology between those with and those without is known as the "digital divide."

MULTIPLE CHOICE QUESTIONS

1. **The third position in the manufacture of cars in the world is occupied by**
   a. USA
   b. France
   c. Japan
   d. Germany
2. **In terms of the monetary value, the largest ship-builder of the world is**
   a. Japan
   b. South Korea
   c. China
   d. U.S.A.
3. **HDI refers to**
   a. Human Development Index
   b. Human Developed Index
   c. Humanity Development Index
   d. None of the above
4. **Economic geography studies how economic activity is distributed geographically as well as the factors and mechanisms that affect it.**
   a. True
   b. False
5. **Which country is the largest producer of raw jute?**
   a. Bangladesh
   b. Pakistan
   c. India
   d. Myanmar
6. **The river also known as Tsangpo in Tibet is**
   a. Indus
   b. Sutlej
   c. Brahmaputra
   d. Teesta
7. **Primary economic activity include**
   a. mining

b. forestry
c. grazing
d. fishing
e. all of the above

8. **Primary activity workers are known as "red-collar workers" because they spend the majority of their workdays outside.**
a. True
b. False

9. **The largest iron-ore producing region in the U.S.A. is**
a. Albama region
b. Lake Superior region
c. North East Appalachian region
d. Sierra Nevada region

REVIEW QUESTIONS

1. Explain when and how Economic geography began?
2. Define Economic Geography and give definitions of Geographers.
3. Explain the nature of Economic Geography.
4. Write a note on the scope of Economic geography
5. Critically evaluate the proposition that economic globalization leads to the 'end of geography'.
6. Explain the key differences between economic orthodoxy and an economic-geographical approach to the economy.
7. What are the two main forms of economic distribution? What is the difference between them?
8. What is the distinguishing characteristic of institutions in the public purpose sphere?

Answer to Multiple Choice Questions

1. (d)	2. (b)	3. (a)	4. (a)	5. (a)
6. (c)	7. (e)	8. (a)	9. (b)	

REFERENCES

1. Aoyama, Y., James, T. M., & Susan, H., (2010). *Key Concepts in Economic Geography*. London: SAGE.

2. Barnes, T. J., Jamie, P., Eric, S., & Adam, T., (2004). *Reading Economic Geography*. Oxford, UK: Blackwell Publishing Ltd. http://dx.doi.org/10.1002/9780470755716.
3. Clark, G. L., Maryann, P. F., & Meric, S. G., (2000). *The Oxford Handbook of Economic Geography*. New York: Oxford University Press.
4. Coe, N. M., Philip, F. K., & Wai-Chung, Y. H., (2007). *Economic Geography: A Contemporary Introduction*. Malden, MA: Blackwell.
5. Dicken, P., (2011). *Global Shift: Mapping the Changing Contours of the World Economy* (6th edn.). New York: Guilford.
6. Barnes, T. J., (2004). *Reading Economic Geography*. Malden, MA: Blackwell Pub.
7. Stutz, F. P., & Barney, W., (2011). *The World Economy: Geography, Business and Development* (6th edn.). Upper Saddle River, NJ: Pearson Education.
8. Vernon, H. J., (2005). *New Economic Geography*. Cheltenham, UK: Edward Elgar Pub.
9. Wood, A., (2010). *Economic Geography*. Milton Park, Abingdon, Oxon: Routledge.

CHAPTER

2 Resources

LEARNING OBJECTIVES

After studying this chapter, you will be able to:

- Discuss land resource
- Explain water resources
- Focus on mineral resources
- Describe energy resources
- Understand nuclear energy resources

"Sustainable development requires human ingenuity. People are the most important resource."

—*Dan Shechtman*

INTRODUCTORY EXAMPLE

Resource Conservation

Every resource created by humans relies on one or more natural resources. Before conserving any natural resource, it's crucial to acknowledge their importance. We should aim to preserve them for future generations while utilizing them responsibly in our lifetimes.

For example, a water shortage in society is imminent if we don't start conserving water. Similarly, we'll deplete wood resources if we continue clearing forests, hindering construction and fuel needs. This would also lead to increased carbon dioxide levels, posing a threat to human life. Protecting resources is vital for human well-being.

We can help conserve natural resources by using them thoughtfully and allowing them to replenish naturally. Sustainable development, ensuring the balanced use of resources, is another key focus. By working together, we can create a planet that is abundant in resources and healthy.

INTRODUCTION

A resource is anything that has value and is needed to achieve a certain goal. Any material that is present in our surroundings and aids in meeting our needs and desires is referred to as a resource. With time and evolving technology, an item turns into a resource.

Any nation's economic resources that are found on its territory are referred to as domestic resources. Despite their existence and potential, these resources are not used in concert. This consequently causes a discrepancy between those resources' estimated potential and actual utilization.

Resources have three common features:

- Utility: Resources are used to satisfy the consumer.
- Limited availability: Resources are inadequately available against demand.
- Potential for depletion: Resources are gradually reduced as they are used.

A country's economy relies mainly on its resources. Industry and agriculture cannot progress without land, water, forests, air, and minerals. These resources, including air, water, forests, and other natural environments essential for human survival and progress, are utilized. By making use of the resources provided by nature, humans have constructed their homes, buildings, industries, and modes of transportation and communication. Alongside natural resources, man-made resources are also crucial and valuable for development.

2.1. LAND RESOURCE

Learning Objectives

- Understand land-use patterns and changes
- Identify problems of land resources:
- Focus on land use planning and policy

Land resources are those that come from the land itself. These include raw materials such as coal, bauxite, gold, and other minerals; subterranean water; and agricultural land that naturally fertilizes the products sown.

- The more a country is able to locate and use the raw materials, the more the industrialization of that country, area, etc.

The term "land resource" refers to any land that can be used for development, such as non-agricultural areas for buildings or township expansion. Natural resources, also known as "land" or "raw materials" in the business sense, are found naturally in environments that have been largely untouched by human activity.

- A natural resource is often characterized by amounts of biodiversity existent in various ecosystems.
- Natural resources are derived from the environment. Many of them are essential for our survival while others are used to satisfy our desires.
- Natural resources can be classified in various ways.
- Underground geological resources (oil, gas, ores, precious metals), and
- While some nations recognize deeper geohydrological resources, such as confined aquifers, as part of individual land ownership with the subsequent right to exploit or sell them, they are not considered part of the natural land unit.

2.1.1. Land-Use Pattern and Changes

The way land is used and arranged determines its land use pattern. This pattern is influenced by various factors like climate, soil, population density, and technical and socioeconomic features. The main land use categories include forests, wastelands, agriculture, and human habitation. In modern society, the most heavily utilized land is allocated for housing, industries, roads, and human settlement.

Land left unused for three to five years is known as fallow land, while abandoned or underutilized urban spaces like playgrounds or vacant lots are considered wasteland.

Areas that cannot support vegetation and remain unused by humans are also classified as wasteland. Fallow land retains some fertility even after years of being uncultivated, making it suitable for crops. Challenges related to land resources stem from physical, economic, and political factors.

2.1.1.1. Changes in Land Use Patterns

There are two categories of anthropogenic changes in land use patterns: direct changes and indirect changes. The type of economic activity in a location greatly influences how the land is used. Unlike many other natural resources, the land's area remains stable, but economic activity fluctuates over time.

Understanding the three distinct categories of economic shifts and their impact on land use is crucial now. India's economy has undergone significant changes in the last 50 years, leading to alterations in land use across the nation.

While four categories experienced growth, the remaining four saw a decrease. The percentages of land used for non-agricultural purposes, fallow area currently in use, net sown area, and forest coverage have all increased.

There are two types of land use change: direct anthropogenic changes (caused by humans) and indirect changes. Examples of anthropogenic changes include agriculture, urbanization, afforestation, and deforestation. Indirect changes involve variations in temperature or carbon dioxide levels that lead to vegetation changes.

There has been a decline in four types of lands: fallow, arable wastelands, areas with pastures and tree crops, and barren and barren lands. Major changes in land use patterns have occurred in India both spatially and chronologically. Even though land use has changed dramatically across the nation, the Green Revolution has had a particularly large impact on agricultural land use. Land use change is responsible for about 25% of all emissions worldwide.

- This can affect the amount of carbon dioxide in the atmosphere, which contributes to the trend of global warming.
- The annual cost of land degradation is $10.6 trillion.
- According to a research analysis conducted by the Food and Agriculture Organization of the United Nations, more than 500 million hectares of new agricultural land will be needed to meet the world's food needs by the year 2050.

Keyword

Biodiversity is the variety of life found in a place on Earth or, often, the total variety of life on Earth.

Remember

The use of land is determined both by physical factors such as topography, climate, soil types, as well as human factors such as population density, technological capability, culture, traditions, etc.

Question

What are the Changes in Land Use Patterns?

Answer

Direct anthropogenic changes and indirect changes are the two types of changes in land use patterns. Since the economy has changed significantly in the past few decades, there has been a major impact on land usage in the country.

Keyword

Climate change is the significant variation of average weather conditions becoming, for example, warmer, wetter, or drier—over several decades or longer.

2.1.2. Problems of Land Resources

As air and water are essential to humankind, so too is land. Food insecurity, increased food costs, climate change, environmental risks, biodiversity loss, and ecosystem services are all consequences of land degradation, which is the deterioration or loss of the soil's productive capacity for the present and future. The global productivity of croplands and rangelands has dramatically declined due to land degradation, which is occurring at an alarming rate.

One of the most urgent environmental issues facing the planet is land degradation, which will only get worse if immediate corrective action is not taken. Approximately 25% of the Earth's surface has been degraded worldwide. Degradation of land releases carbon from the soil and nitrous oxide into the atmosphere, making it one of the main causes of climate change. Scientists recently announced that unsustainable agricultural practices were the main cause of the annual loss of 24 billion tons of fertile soil. Ninety-five percent of Earth's land area could be degraded by 2050 if current trends continue.

Three and a half billion people worldwide are impacted by land degradation, with smallholder farmers, rural communities, and the extremely poor being the most affected. The demand for agricultural products, such as food, feed, fuel, and fiber, is expected to rise as the world's population rises by approximately 35% to 9.7 billion by 2050. However, there are other factors as well that are putting more strain on the world's land supply, like natural phenomena such as climate variability and extreme weather events, and agricultural production systems that are becoming less resilient due to biodiversity loss. The resilience of agro-ecosystems and the stability of food production systems are threatened by climate change, which intensifies variations in agricultural yields and revenue. The driest regions of the world have especially serious issues. Two billion people live in desert landscapes, which make up about 40% of the planet's land area. Women and children are particularly vulnerable to the effects of drought and land degradation in developing nations, where the bulk of people who depend on drylands reside.

2.1.3. Land Use Planning and Policy

Land is a vital resource for the impoverished in rural and urban areas, serving as a crucial basis for their social and economic advancement. Equitable access to land and secure property rights can empower those in need in rural and urban areas, helping them adapt to challenges posed by current globalization trends. Various shortcomings in land-use policy, administration, and management negatively affect land markets' efficiency, impacting the nation's economic growth and equity. The text highlights how community and governmental efforts can reduce insecurity. By implementing these strategies, land can be utilized more effectively, crucial for promoting growth and reducing poverty.

2.1.1.2. Land Use Policy

A growing number of non-policy factors, including globalization, demographic shifts, and climate change, as well as changes in agricultural, agri-environmental, land use, and regional policies, influence decisions about land use and management. Alternative means of influencing agricultural land use are provided by land-use policy. Traditionally, land-use policy hasn't been thought to be important for agriculture.

Most land-use policies in most countries use the government's regulatory power to force landowners to comply with specific regulations, though some use financial incentives to change the relative returns to farmers from agriculture and other uses. This makes the majority of land-use policies non-market-based.

Agricultural land use is influenced by land-use policy in two different ways. The first and most popular approach is to either encourage particular behaviors among farmers or impose limitations on their behavior. The second category of policies aims to impact decisions about the larger-scale use of land, i.e., whether to convert land to agricultural use or take it out of farming.

Several goals concerning the distribution and security of land rights, land use and management, and access to land, including the tenure arrangements under which it is held, are the focus of land policy. In essence, a land-use policy is the government's opinion of the course to be followed on important land-use-related issues and the suggested distribution of the country's land resources over a

Did you know?

India supports approximately 16% of the world's population and 20% of its livestock in 2.5% of its geographical area. This pressure on land has led to its deterioration – soil erosion, waterlogging, salinization, nutrient depletion, lowering of groundwater tables, and soil pollution – largely caused by human interventions.

predetermined amount of time. There are two parts to it: production and conservation. A strong national land-use policy that addresses all land uses is an essential component of the enabling environment.

To attain the policy goal of sustainable production and preservation of natural resources, governments ought to actively support land use practices that are both aesthetically pleasing to the public and environmentally sustainable in terms of their effects on land resources. The chances of accomplishing this are significantly increased if the national land-use policies are developed via an integrated, participatory, and iterative process.

Keyword

Economic Development includes programs, policies or activities that seek to improve the economic well-being and quality of life for a community.

Demands for food, water, energy, and raw materials are being driven by population growth, economic development, and urbanization. Additionally, the constant shift in the human diet from cereal to animal products, which requires more land and water resources, and the recent shift to biofuels increase the need for farm production, which has an impact on land uses.

Like any other type of agriculture, increased biofuel production has the potential to endanger biodiversity, water and land resources, and other factors. To reduce any potential negative effects, the right policies must be put in place. The term "land-use policy" pertains to the collection of regulations and guidelines that directly impact the utilization of agricultural land, regardless of whether they are established by national or sub-national authorities. When viewed through the lens of rural development, policy considerations regarding alterations in farmland use can be categorized into the following five distinct aspects:

- Adverse environmental impacts on landscape provision, wildlife habitat, and the preservation of ecosystems, stemming from the abandonment of farmland in some rural areas of high nature value;
- Knock-on economic effects of the abandonment or long-term retirement of farmland influencing the socio-economic viability of such rural areas;
- Risks to the provision of farmland-based rural amenities, particularly in those rural areas where such amenities are instrumental for their sustainable development;
- concern with the alternative uses of farmland and water in the encouragement of environmentally sustainable rural development and alternative sources of income and employment in rural areas; and
- urban sprawl in cases where farmland is lost to urban uses. One role of policies is to narrow the divergence between privately and socially desirable outcomes.

2.1.1.3. Land Use Planning

An efficient way to ensure that activities near airports are suitable for aviation is through land-use planning and management. By implementing land-use zoning around airports, the primary objective is to reduce the population's exposure to aircraft noise. A crucial tool for preventing additional residential development near airports from negating the benefits of quieter, more modern aircraft is compatible land-use planning and management (Figure 2.1).

Figure 2.1. Gravel quarry at Ottenhoefenim Schwarzwald, Germany. Residential and industrial usage abut with no buffer zone.

Source: By Robbie Morrison – Own work, CC BY 4.0, https://en.wikipedia.org/wiki/Land-use_planning#/media/File:Gravel_quarry_at_Ottenhoefen_im_Schwarzwald.jpg.

Cities and communities would be in disarray without an appropriate land-use plan. There would be inefficient transportation. Residential areas, streams, and air quality would all be contaminated by industrial plants. Cities would not be feasible. Localities would be in danger. Economy stalling would occur. The region would become unusable for future generations due to resource waste and environmental damage.

Since ancient times, there has been land planning. As people started to live in bigger and bigger groups instead of as lone nomads and small tribes, and as villages grew into bigger metropolitan areas, land planning became crucial to day-to-day existence.

Relics of land zoning and development can still be observed in the ruins of Egypt, Greece, Rome, Mexico, and other places. The reason for this is that our forefathers realized how advantageous it was to have designated places for living, working, shopping, disposing of waste, and burying the dead.

The importance of land-use planning for urban public health, however, was not fully appreciated until the 1900s. Insufficient planning in big industrial cities in the U.S. Research from cities like Chicago, New York, and San Francisco showed that living close to industrial areas promoted the spread of illnesses like influenza and tuberculosis as well as an increase in crime.

Usability is the one word that best describes why creating a proper land use plan is essential. Land planning is absolutely necessary for the orderly development and use of the area because the community uses the land and its resources.

a. Importance of Land Use Planning

People are becoming more conscious of how we affect the environment in today's society. We recognize that there is a carbon footprint associated with each of our actions. Our increased awareness indicates that we are becoming more aware of our work and living habits, how they interact, and how to live sustainably.

To shorten commutes and promote low-emission modes of transportation like walking and biking, there is an increasing demand for walkable communities, mixed-use developments, and green spaces where people can live, work, and play. Creating a sustainable environment is the sole goal of these initiatives. When land use planning is done correctly, it takes into account the needs of the present and projects those into potential future needs. This results in a long-term, sustainable vision that permits the development of land to meet those needs.

Here are a few reasons land use planning is so important:

- Reduces pollutants
- Improves air and water quality
- Prevents flooding
- Promotes safe transportation and routes
- Reduces noise and light pollution
- Keeps the environment ecologically balanced
- Protects potentially endangered species
- Mitigates disaster risks
- Safeguards public health and minimizes hazards
- Maintains and improves the fiscal health of a community
- Preserves open space
- Manages growth

b. The Land Development Planning Process

From the time of land acquisition to project completion, years may pass. The initial phase of any project is land development planning.

Depending on the scope of the project and the location, several steps may be involved, including:

- Market analysis and feasibility studies
- Land acquisition or securing the rights to purchase the land
- Environmental assessments
- Surveys
- Developing plans for the site, development, and any buildings
- Acquiring permits
- Financing
- Applying for rezoning* (*In some cases)

All of these things happen before a building project ever gets off the ground. The project plan, infrastructure development, construction, permits and inspections, final walkthrough, and design iterations are the last stages.

c. How To Make a Land Development Plan

Utilize the expertise and skills of an experienced land-use planning company to create a comprehensive land development plan. Select a land planning company in Utah that is knowledgeable about the nuances and various laws that may be relevant to your project.

Land development plans encompass the following considerations:

- What site constraints and opportunities exist? (Access to utilities, easements, zoning, encroachments, etc.)
- Are there environmental, hazardous, archaeological, wetlands, water rights, threatened or endangered species, or other government-related constraints?
- The project's financing and strategies for potential delays or cost increases.
- The marketability of the project

- Impact studies (traffic, stormwater run-off, water, etc.)
- Community master plans
- Timeline of reviews and permits and a flow of what's needed
- Cost breakdown of each element
- Site plan
- Construction sequencing
- Logistics management plan

d. The Common Types of Land Use Planning

Various forms of land use planning exist in urban areas and other communities. However, the most common types of land use planning are the following five types:

- Recreational: Land designated for recreational purposes includes areas like playgrounds, golf courses, open spaces, equestrian centers, parks, fishing ponds, and swimming pools.
- Transport: Airports, light rail, buses, trains, interstates, and other land required for transportation are all located on this type of property. This zoning category includes structures used for municipal transportation, such as airports, subway stations, bus terminals, and other buildings.
- Agricultural: A farm's buildings, fields for animals, crop cultivation and harvesting, and other farming operations may be permitted on land designated for agricultural use. Local laws may specify the number, kind, and type of animals that can be kept, as well as the types of crops that can be grown and the methods for controlling pests. Water rights and restrictions are frequently attached to agricultural use.
- Residential: Residential zoning varies according to the types and densities of houses that can be constructed in the designated area. These could be low-density single-family homes, townhomes, and condominiums with a moderate density, multi-family apartment buildings with a high density, or multi-use complexes. A distinct zoning category applies to multi-use or mixed-use developments, which combine residential, commercial, and recreational space usage.
- Commercial: Restaurants, retail stores, office buildings, and other commercial uses are permitted under commercial zoning. Zoning ordinances have distinct sub-designations that impose limitations on the kinds of commercial activity that are permitted due to the vast range of possible commercial uses.

Public-use buildings, such as fire stations, police stations, and hospitals, as well as industrial and tourism-related buildings, fall under additional zoning categories and land-use planning factors. Community land-use planning depends critically on emergency resources and their prompt availability.

e. The Factors Affecting Land Use Planning

There are so many variables that impact land use planning that they would fill an entire college textbook and are too numerous to list here. However, the following are the main variables that influence land use planning.

- Sustainability
- Energy
- Social Equity
- Economy
- Livability

- Transportation
- Housing
- Environment
- Health
- Food
- Recovery
- Climate Change

Future considerations are something that land-use planning firms have to keep in mind when creating community plans.

- Changing demographics
- Lifestyles
- Potential disasters, hazards, and how to mitigate them (hurricanes, earthquakes, floods, etc.)
- Transportation types and technological advancements
- Housing improvements and advancements
- Energy conservation
- Green technology and improvements

Did you get it?

1. **How do land use changes affect the environment?**
2. **What is causing the changes in land use patterns?**
3. **Why is land use planning important?**

2.2. WATER RESOURCES

Learning Objectives

- Understand the extent of water resources
- Focus on water resource potential
- Identify utilization and related problems
- Know water disputes
- Perform water management, conservation, and development

For all living things, water is one of the most essential resources. Even though water is a renewable resource, there are still many places in the world where there is a severe shortage of good water. We require water for many things, including food production, hygiene, energy production, fire control, and—above all—survival.

2.2.1. Extent of Water Resources

Water resources in the country are extremely valuable. These resources include streams, lakes, wetlands, coastal waters, groundwater, and other bodies of water; the ecosystems that they support; and the uses by humans, including recreational activities, drinking water, and fish supply. Ecosystems, human uses, and the overall health and sustainability of the hydrologic cycle depend on the extent (amount and distribution) and condition (physical, chemical, and biological) of water resources.

The mission of the Environmental Protection Agency (EPA) is fundamentally linked to the protection of water resources, as their quantity and quality can impact ecosystems, human health, and essential environmental processes. Together with other federal, state, local, and tribal government agencies, the EPA collaborates with others who share an interest in the quantity and quality of water resources.

2.2.2. Water Resource Potential

Water resources are under unprecedented strain in the majority of countries today. The world's population is expanding quickly, and projections indicate that by 2030, there will be a 40% gap between the world's predicted water demand and supply due to current practices. Furthermore, some of the greatest threats to the prosperity and stability of the world are thought to be chronic water scarcity, hydrological uncertainty, and extreme weather events like floods and droughts. Recognition of the part drought and water scarcity play in escalating vulnerability and conflict is growing. A 50% increase in agricultural production—which currently uses 70% of the resource—and a 15% increase in water withdrawals are needed to feed 10 billion people by 2050. In addition to the rising demand, the resource is already in short supply in many regions of the world. Over 40% of the world's population, according to estimates, lives in areas with limited access to water, and this problem affects about ¼ of the global GDP. One in four kids are predicted to reside in regions with acute water scarcity by 2040. For many nations today, ensuring water security is a significant and frequently expanding challenge. The situation will get worse as a result of climate change, which will change hydrological cycles, make water more unpredictable, and increase the frequency and severity of floods and droughts. The 500 million people who live in deltas and the approximately 1 billion people who live in monsoonal basins are particularly at risk. The annual cost of property damage from floods is estimated to be around $120 billion, and the impoverished in rural areas who rely heavily on variable rainfall for their subsistence face various challenges due to droughts.

Water security faces challenges due to the fragmentation of this vital resource. Sixty percent of the world's freshwater flow is shared among 276 transboundary basins across 148 nations. Approximately 2.5 billion people globally depend on groundwater, with 300 aquifer systems being transboundary. Collaboration is key to finding optimal solutions for water resource development and management that benefit all riparian communities. The issues stemming from fragmentation often extend to the national level, emphasizing the need for effective water resource management and services. Countries must address these complex water challenges by managing resources more efficiently.

Given the rising demand, water scarcity, uncertainties, extreme events, and fragmentation issues, clients should consider investing in information management, institutional strengthening, and infrastructure development (both natural and man-made) to enhance water security. To better manage, regulate, and conserve water resources, tools like water pricing, legal frameworks, and incentives are essential. Information systems play a crucial role in hydro-meteorological forecasting, decision-making in uncertain conditions, and resource monitoring. Furthermore, exploring methods to enhance water storage, such as aquifer recharge, and investing in innovative technologies to boost productivity, preserve resources, recycle stormwater and wastewater, and tap into unconventional water sources are crucial. Rapid adoption and application of these advancements will be vital to enhancing water security globally.

2.2.3. Utilization and Related Problems

A shortage of drinkable water for a population is called a water crisis. Drought, starvation, and death are the eventual results of this condition. For those residing in drought-stricken areas and on the African continent, clean drinking water is now considered a luxury. People are seen looking for it by walking miles and spending all day. More depressingly, once they acquire it, they will once more have to contend with the waterborne illnesses that result from it. When basic needs are not met and people have to work hard to obtain them, economic development suffers. Nevertheless, we humans take it for granted and fail to see how crucial water conservation is.

2.2.3.1. Water Availability Use Pattern and Scarcity

Over 97% of the water on Earth is salty, and the remaining 2% is frozen in ice caps, which is where rivers get their water from. Less than 0.7% of the water in the world is suitable for human use. The estimated total volume of water on Earth is $385{,}580 \times 10^{10}$ gallons, meaning that each person has a share of roughly 9×10^{10} gallons. Freshwater supply is undoubtedly limited because saline water makes up roughly 97% of all water.

Freshwater available for human use is currently between 10,000 and 12,000 cubic kilometers per year worldwide. Approximately 4,000 km^3 were removed from this supply by 2000 people. Approximately 50% of the water extracted evaporates, gets transpired by plants, or becomes contaminated beyond use, making it temporarily unavailable for use by other users. The remaining 50% is put back to use; most urban wastewater is cleaned up and put back into use, and some irrigation water drains back into rivers or replenishes groundwater. Of the water taken out for human use, 65% is used for agriculture, 10% is used for domestic purposes (households, public utilities, commercial use, and municipal water systems), 20% is used for industry (primarily the production of electricity), and 5% evaporated from reservoirs. Comparatively speaking, about 70% of the water used for agriculture was used, compared to 14% for household use and 11% for industrial use. The world's water use is largely determined by economic development and population density. As the world's population grows from 6.1 billion to 7.5 billion people by 2050, the World Water Council predicts that total yearly withdrawals will reach more than 5,000 km^3.

Keyword

The hydrologic cycle, also known as the water cycle is a way of describing the material flow of water throughout the Earth.

The point where the combined effects of all users start affecting the quantity or quality of water under the current institutional arrangements to the extent that demand from all sectors, including the environment, cannot be fully met is known as water scarcity. Water scarcity is a major global issue in the twenty-first century. The rate at which water use has increased is more than double that of population growth compared to the previous century. There are increasing areas facing persistent water deficits, with water shortages present on every continent. Around 500 million people are at risk of water scarcity, and approximately 1.2 billion people, or nearly one-fifth of the world's population, live in areas with physical water shortages. In some countries, there is inadequate infrastructure to extract water from aquifers and rivers. About 166 million individuals worldwide, or nearly 25% of the total population, face financial water scarcity (Figure 2.2).

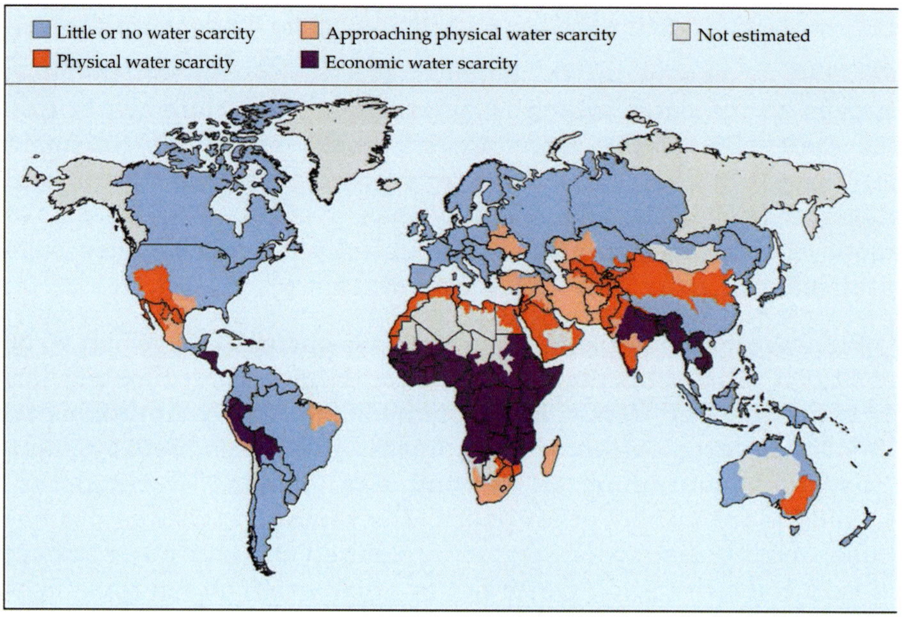

Figure 2.2. Water scarcity regions of the world.

Source: Mancosu, N., Snyder, R. L., Kyriakakis, G., & Spano, D. (2015). Water Scarcity and Future Challenges for Food Production. Water. 7(3), 975–992. https://doi.org/10.3390/w7030975.

Keyword

Groundwater is fresh water located in the subsurface pore space of soil and rocks. It is also water that is flowing within aquifers below the water table.

If there wasn't an uneven distribution of freshwater, there would be enough freshwater on the planet to support seven million people. This is one of the causes of water scarcity. However, a significant amount of water is squandered, contaminated, and managed unsustainably; as a result, we can draw the conclusion that human and natural factors contribute to the current state of water scarcity. Water resources around the world are more affected by climate change. Droughts can occur when evaporation rises in response to temperature increases. Furthermore, glacial ice is melting due to rising temperatures. The primary source of freshwater, glaciers, may vanish by the end of the twenty-first century. The most significant effect of rising temperatures and consequent water scarcity is receding glaciers. According to climate scientists, the hydrologic cycle will be significantly impacted by climate change, and these effects will frequently exacerbate already-existing water shortages. Increased global temperatures will change the patterns of rainfall, making them stronger in some places and weaker in others. They may also increase the frequency and severity of storms in some parts of the world. Let's sum up by saying that many of the water-management issues discussed in this unit will probably become even more complicated as a result of climate change.

There's a shortage of water for about 700 million people worldwide, spread across 43 countries. Mainly due to climate change, between 75 and 250 million people in Africa are expected to be part of the nearly half of the world's population living in regions facing severe water stress by 2030. The number might increase in Sub-Saharan Africa. To tackle the global water crisis, the UN has launched an awareness campaign through World Water Day to combat desertification.

In many parts of the world, threats to water resources come from the expansion of agriculture, damming, diversion, overuse, and pollution. Providing safe drinking water to over 1 billion people facing water scarcity is a major challenge for governments. A significant portion of the population lacks access to clean, uncontaminated water, especially in developing nations, where water contamination remains a concern. Even wealthy nations with advanced water supplies and treatment technologies encounter difficulties. Many regions are now looking to more distant water sources due to overdevelopment, particularly in coastal areas and regions with limited water resources. The average daily water usage per person in a city is 150 gallons, which is also used for manufacturing, sanitation, and cleaning.

Diseases related to water that can be prevented claim millions of lives every year. 1.7 billion people (17% of the world's population) lacked access to clean drinking water in 2002, while 20.6 billion people (42% of the world's population) lacked proper sanitation. A shortage of clean drinking water is a problem for the nations of Africa, Asia, and the Pacific. But the issue still exists in other countries. Adequate sewage treatment services are lacking in many European households. Water use is unevenly distributed; poor communities in both rural and urban areas frequently lack access to clean water and sanitary facilities, and cities frequently receive better service than rural areas.

2.2.3.2. Water Scarcity and Related Problems

The current state of water scarcity in the nations of the world is influenced by numerous factors. Here are some descriptions of them.

a. Depleting Water Resources

Globally, surface and groundwater water resources are disappearing at a startling rate. There are numerous explanations. But there are three main ones. The first is population growth. Freshwater is becoming scarce while its applications in agriculture, industry, domestic use, and drinking are growing. Pollution is the second explanation. Large-scale human activity has contaminated water sources, making it unusable for daily human needs in many places. Climate change is the third reason. The worldwide precipitation pattern has changed due to climate change, and it is now rare in most places. Another reason for future water scarcity is the melting of ice caps and glaciers brought on by global warming. Aquifers are being depleted by human activity faster than they are being recharged. Excessive groundwater pumping can drain neighboring rivers and streams and alter groundwater flow patterns around wells, in addition to depleting aquifers.

b. Water Salinization

Many regions of the world are experiencing an increase in water salinity as a result of various factors. The primary causes of these are extended dry spells, insufficient precipitation, and overuse of surface water. Freshwater resources that turn salty cannot be used for drinking or irrigation. High sodium levels make dry soils compact and harden, which hinders their capacity to absorb water. Saline water is also toxic to plants. In contrast to seawater, which has a salinity of roughly 35,000 mg/l, irrigation water becomes toxic to most plants at concentrations greater than 1,300 mg/liter.

c. Water Pollution

Keyword

Pollution is the introduction of harmful materials into the environment. These harmful materials are called pollutants.

Water can become contaminated and unfit for human use by a variety of substances. Primary standards for pollutants in drinking water include:

- Microorganisms such as cryptosporidium, giardia, and fecal coliform bacteria
- Disinfectants and water disinfection byproducts including chlorine, bromate, and chlorite
- Inorganic chemicals such as arsenic, cadmium, lead, and mercury
- Organic chemicals such as benzene, dioxin, and vinyl chloride
- Radionuclides including uranium and radium
- Surface water – Water in rivers, lakes, and wells is severely contaminated, mostly in developing nations, making the water unfit for human use and adding to the already dire shortage. The main causes of water pollution are industrialization and development across all domains.

d. Water-Borne Diseases

Diseases like cholera, typhoid, and dysentery that are transmitted by tainted water or a shortage of water for hygiene cause the deaths of over two million people annually. Drinking water contaminated with bacteria or viruses that spread through human or animal waste can result in cholera, typhoid fever, and dysentery. Lack of clean water for washing is the main cause of skin and eye infections. The developing world is primarily responsible for this issue.

Question

How can we prevent water scarcity?

Answer

While some human activities have exacerbated the water crisis, humans have also developed technologies to better acquire or conserve freshwater. Solutions to addressing water shortages include dams and reservoirs, rainwater harvesting, aqueducts, desalination, water reuse, and water conservation.

2.2.4. Water Disputes

Water is crucial for human survival. Even though freshwater is abundant, its supply is limited. Due to the increasing global population and economic levels, the demand for freshwater has been rising rapidly.

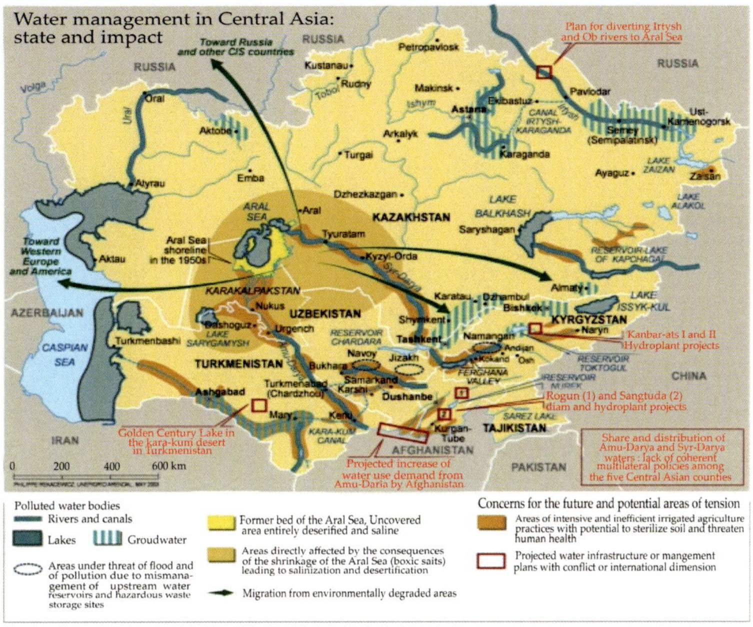

Figure 2.3. Water management in Central Asia.

Source: Peña-Ramos J. A., Bagus, P., & Fursova, D. (2021). Water Conflicts in Central Asia: Some Recommendations on the Non-Conflictual Use of Water. Sustainability. 13(6), 3479. https://doi.org/10.3390/su13063479.

The availability and quality of water are changing regionally and seasonally due to climate change and environmental degradation. While researchers emphasize that conflicts are rarely caused by a lack of water itself, but rather by governance and management issues, competition over water use can lead to conflicts and sometimes violent outbursts (Figure 2.3).

We explore the links between water and conflict through ten case studies from the interactive ECC Factbook in the Editor's Pick. They delve into different efforts to seek peaceful resolutions and analyze the various connections between security and water.

2.2.4.1. Dispute Over Water in the Nile Basin

Among the eleven riparian countries that make up the Nile Basin, there is a significant amount of conflict over rights to and access to water resources. Established in 1999 with support from major donor institutions and nine out of ten riparian countries, the Nile Basin Initiative (NBI) has made some progress in enhancing cooperation. However, since 2007, disagreements over interests between countries upstream and downstream have stalled negotiations, pitting Egypt against upstream riparians, particularly Ethiopia, and to a lesser extent, Sudan. A framework agreement that may eventually pave the way for a more comprehensive agreement was reached in 2015 as a result of trilateral talks between these nations regarding a significant dam being built in Ethiopia.

2.2.4.2. Water Shortages and Public Discontent in Yemen

The availability of water in Yemen is drastically decreasing due to serious mismanagement. The causes of the unequal distribution of effects on the populace are primarily corruption and nepotism. The lack of water has made the underprivileged more and more irate, contributing to Yemen's political and security issues.

2.2.4.3. Turkey, Syria and Iraq: Conflict over the Euphrates-Tigris

Iran owns a portion of the Tigris basin, while Turkey, Syria, and Iraq share the Euphrates-Tigris basin. The basin's relations have been strained since the 1960s due to political tensions between the countries and unilateral irrigation plans that change the rivers' flows. The three governments have been unable to effectively co-manage the rivers in the basin due to disagreements. A formal agreement on managing the basin waters has not yet been reached, despite renewed cooperation efforts in the 2000s.

2.2.4.4. Transboundary Water Disputes between Afghanistan and Iran

Iran has shown concern about Afghanistan's plans to utilize the Helmand River and the Harirud River for post-conflict development and reconstruction. The Iranian government sees the development of agriculture and construction of dams in Afghanistan as potential threats to the water supply security in its eastern and northeastern regions. Despite cooperative efforts, there hasn't been a significant breakthrough due to an ineffective water treaty. Afghanistan's reluctance to engage in water discussions, coupled with Iran's conflicting actions, has complicated the resolution of the two countries' water disputes.

2.2.4.5. Dam Projects and Disputes in the Mekong River Basin

The construction of dams to produce hydropower is rapidly expanding throughout the Mekong basin, particularly in China and Laos. Diplomatic tensions have arisen as a result, with countries downstream of the dams fearing worse effects, such as increased flooding or seasonal water shortages. China's unwillingness to become a full member of the Mekong River Commission (MRC) and the commission's lack of enforcement authority have limited the MRC's ability to effectively resolve these tensions thus far. China is attempting to interact with downstream riparians in place of joining the MRC by putting forth alternate institutional mechanisms and providing support for the construction of dams in the Lower Mekong basin. However, current dam-building activities may continue to be a destabilizing factor in the Mekong River Basin in the absence of more formalized cooperation, particularly between China and the lower riparians.

2.2.4.6. Dispute over Water in the Cauvery Basin in India

Amid drier climate conditions, the long-standing dispute between Tamil Nadu and Karnataka, two Indian states, over water from the Cauvery River, has recently surfaced. In the wake of decisions to change the distribution of water between the two states, there will likely be violent protests in addition to legal disputes.

2.2.4.7. Droughts, Livestock Prices, and Armed Conflict in Somalia

In Somalia, the livelihoods of pastoral people are severely impacted by frequent droughts. Reduced livestock prices and declining rural incomes are the results of herders selling more animals during droughts than they would normally do. The combination of pervasive poverty and a dearth of viable economic options encourages people to engage in illegal activities and join armed groups like Al Shabaab, which pay their fighters in cash and other benefits. The extreme Islamist group is thought to have grown in numbers, particularly in 2011 during the record drought.

2.2.4.8. Turkey-Armenia: Water Cooperation Despite Tensions

The case of Turkey and Armenia is a well-known illustration of how two co-riparians can set aside their differences, cooperate for their shared benefit, and share transboundary waters fairly.

2.2.4.9. Security Implications of Growing Water Scarcity in Egypt

Egypt is currently using more water than it is able to produce internally through renewable resources, primarily the fresh water inflows from the Nile. Future increases in Egypt's water stress are anticipated due to the country's rapidly expanding population, warming temperatures, and growing water consumption. Egypt's economy will be severely strained and more susceptible to renewed internal conflict if the country's growing freshwater shortage is not appropriately addressed. It also runs the risk of exerting more pressure on Egypt's diplomatic ties with other states bordering the Nile.

At least nine people lost their lives in the so-called "Water War of Cochabamba," which broke out in 2000 when the city of Cochabamba's drinking water was privatized. The protests were violent. Water access eventually obtained new legal support after the city's water was renationalized. Still, Cochabamba faces severe challenges from overuse, declining water supplies brought on by climate change, and inadequate technology.

2.2.5. Water Management, Conservation, and Development

Water is necessary for our everyday existence. There is always room for conservation when using water, whether inside or outside your house or place of business. Since fresh water is a finite resource, water conservation is crucial for the ecosystem. The province's water resources are under increasing pressure due to factors like population growth, industry expansion, increased development activity, and possible climate change effects. Water conservation is a smart move whether you use a private or public home water supply.

Here are some reasons why you might want to use less water:

- Save money on your power bills by using less energy to heat and pump water.
- Delay or prevent the expansion of costly water and wastewater

treatment plants in your community, which can save money on taxes.

- Reduce water shortage frequency and impacts
- Environmental benefits

2.2.5.1. *How to Use Less Water*

The main sources of indoor household water use are faucets, leaks, baths, showers, washing machines, and toilets. Water can be used extensively outside for purposes like recreational activities, driveway cleaning, and lawn irrigation. The steps you can take to conserve water are quite easy. There are several methods you can take to reduce your water usage and support conservation efforts.

1. Check for leaks: Ensuring you are not using excessive water requires regular leak checks. The most common source of leaks in plumbing is toilets.
2. Install water-saving devices: Incorporating water-saving devices into your home or business can aid in water conservation. Consider installing:
    - Water faucet aerators
    - Low-flow showerheads
    - Toilet dams
3. Other simple conservation tips

Conserving Water in the Bathroom: Most indoor water use occurs in the bathroom. Find out how to check for leaky toilets and get additional water-saving advice.

- Install water efficient toilets and fixtures.
- When washing, brushing your teeth or shaving, never let the tap run continuously.
- Turn off the taps tightly (but gently) so they do not drip.
- Take shorter showers.
- Fix leaky faucets and toilets. Even the smallest drip from a worn washer can waste 75 or more liters a day. Larger leaks can waste hundreds.
- Use dye test tablets (found at home improvement stores) or a few drops of food coloring in the toilet tank to check for leaks. Give it ten to fifteen minutes before flushing. A leak is present if the color of the bowl water changes.

Conservation in the kitchen:

- Take foods out of the freezer early to allow plenty of time to thaw rather than running water over them.
- Keep water in the fridge so you don't have to run the tap to get cold water.
- Completely fill the dishwasher before you turn it on. It can use 35–45 liters per cycle.
- If you wash dishes by hand, don't leave the water running for rinsing.
- Don't let the faucet run while you clean vegetables. Just rinse them in a stoppered sink or a pan of clean water.
- Keep cooking fats, oils, and grease out of the drain. They can stick to and block the inside of pipes, which can lead to sewer backups and basement flooding.

Conservation in the laundry room:

- Purchase a high-efficiency washing machine.
- Try to only do laundry when the machine is full.

- If you need to wash smaller loads, be sure to adjust your washer's water level settings accordingly.

Conservation outdoors:

- Deep-soak your lawn, long enough for the moisture to soak down to the roots where it will do the most good. A light sprinkling can evaporate quickly and tends to encourage shallow root systems.
- Water in the morning to eliminate evaporation.
- Adjust sprinklers to water lawns and not driveways or the sidewalk.
- Avoid watering on windy days.
- Put a layer of mulch around trees and plants. Mulch will slow the evaporation of moisture and discourage weed growth too.
- Use a broom, not a hose, to clean driveways and sidewalks.
- When washing your car, rather than using a running hose, consider using a bucket, a sponge, and a hose with a trigger nozzle.
- Maintain a drip-free environment by regularly inspecting garden hoses and connections. Leaks outside the house can be just as wasteful as leaks inside, even though they might not seem as bad because they are hidden.

a. Conservation During Winter

Summertime is usually a time when we use more water to try and keep up with the demands of flowers, vegetable gardens, and lawns. For example, in contrast to other provinces, Newfoundland and Labrador use more water in the winter. The primary cause of this is the possibility of frozen water pipes in cold weather. These are a few strategies for winter water conservation.

- Drip faucets instead of running water to prevent your pipes from freezing.
- Collect water in a bucket for flushing the toilet. Showers take longer to heat up in the winter so take advantage of the first few minutes of cold water.
- Wrap all exposed pipes and insulate hot water pipes to help protect them against bursting.
- If your pipes do burst, know where your shut-off valve is to prevent excessive damage and decrease water usage.
- Call in a plumber after the first thaw to check your pipes for damage and leaks.

Keyword

Population growth is the increase in the number of people in a population or dispersed group.

- Do not use your hose to melt snowbanks. This is an unnecessary waste of water.

It's critical to pay attention to how much water you use, no matter the season. Make sure you adhere to the water-saving tips for the kitchen, laundry room, and bathrooms all year long to avoid using too much water in any of the seasons.

b. Water Efficiency for Business

By reducing water consumption you can:

- Lower operating costs.
- Enhance your public image by demonstrating your commitment to the environment.
- Increase your competitiveness by saving money and attracting new loyal customers.

Did you get it?

1. **What are the potential sources of water?**
2. **What are the potential uses of water?**

2.3. MINERAL RESOURCES

Learning Objectives

- Identify categories of mineral resources
- Define metallic resources
- Explain non-metallic minerals
- Focus on the uses of minerals
- Understand the conservation of mineral resources

The essential building block for socioeconomic development is mineral resources. According to statistics, mineral resources account for more than 95% of the energy used by humans, 80% of industrial raw materials, and 70% of raw materials used in agriculture. Any pure inorganic material found naturally in the crust of the earth is referred to as a mineral. There are over 2,000 known minerals, the majority of which are inorganic and result from different elemental combinations. But organic materials, which are made up of individual elements like sulfur, gold, silver, and diamond, make up a very small percentage of the earth's crust.

2.3.1. Categories of Mineral Resources

When we talk about mineral resources we generally classify mineral resources into two categories:

- Metallic resources; and
- Non-metallic resources.

2.3.1.1. Metallic Resources

Materials such as gold, silver, tin, lead, copper, and so forth are considered metallic resources. While some of these resources are relatively easy to find and extract, others require specialized knowledge and the application of various sciences, and some require unnecessary processing.

Classification of metallic resources

- Ferrous Minerals: Iron makes up the majority of the ore in this metallic mineral component. This component makes up 75% of the total ferrous mineral production. Examples are nickel, chromium, iron ore, etc.

Did you know?

Mineral resources and eras of human technological advancement are intrinsically linked. Early human civilizations are divided into the Stone Age, the Bronze Age, and the Iron Age. You may have also heard our current information technology era described as the Silicon Age.

- Non-Ferrous Minerals: Iron is not the dominant mineral in this component of metallic resources, as the name implies. The remaining 25% of the metallic minerals are made up of these. These materials generally have higher corrosion resistance.

2.3.1.1. Non-Metallic Minerals

The primary component of non-metallic resources is carbon. Natural gas, coal, and petroleum, etc. all fall into this group, and the production of all natural resources is made up of these resources to the tune of about 87%. The remaining portion is split between non-metallic and metallic resources, with each gaining roughly 6–7%. Two additional categories can be used to further classify non-metallic minerals.

- Organic Minerals: Organic minerals are defined as those that are comprised of carbon and hydrogen. This includes natural gas, coal, and petroleum. China mines half of the world's coal, while India mines the remaining 10%.
- Inorganic Minerals: Resources classified as inorganic are those derived from non-metals that lack a carbon-hydrogen bond. Minerals like mica, graphite, and limestone fall into this group. The lines separating organic and inorganic minerals, however, are blurry and appear to overlap in some areas.

Remember

Mineral resources provide us with raw building materials for nearly most objects around us, as well as food, energy, or raw materials that enhance our health and agriculture.

2.3.2. Uses of Minerals

The utilization of mineral deposits is determined by their characteristics. While some nations lack mineral deposits, others have abundant ones. The properties of minerals determine the greatest use for them. For example, aluminum finds application in the automobile, shipping, and aircraft industries due to its lightweight, strength, and durability. Minerals are employed in nearly every industry. The jewelry industry uses metals such as gold, silver, and platinum. Copper is utilized in wire and pipe production, as well as the coin industry. The computer industry uses silicon that is extracted from quartz.

Fireworks get their color from mineral components. Copper yields blues, zinc yields whitish shades, barium produces glossy greens, and strontium produces dark reds. Combining elements yields a wide range of colors: copper and strontium produce lavender blue; strontium and sodium produce bright orange; and alloys of titanium, zirconium, and magnesium produce silvery white.

2.3.3. Conservation of Mineral Resources

A mere 1% of all the minerals found in the earth's crust make up the total volume of consumable mineral resources. But because of how quickly these non-renewable mineral resources are being used up, they will run out. The following actions can be taken to conserve minerals:

- Use of minerals in a planned and sustainable manner.
- Recycling of metals
- Use of alternative renewable substitutes.
- Technology should be improved to use the low-grade ores profitably.

Any mineral will typically exist as fully formed crystals and be handled according to the type of crystal. You may be familiar with some common names, but a comprehensive nomenclature has emerged to classify crystal types. Different characteristics help identify different minerals. These characteristics might not differentiate certain minerals sufficiently to help with detection. Furthermore, they are exclusive to specific minerals.

Question

Why is the conservation of minerals a need of the earth?

Answer

If they get depleted, it might take thousands of years for them to be replenished. Humans are dependent on minerals to satisfy their needs. A wide variety of objects that people use in their daily lives are made from minerals. Thus, it is the need of the hour to preserve minerals.

Did you get it?

1. **Differentiate between metallic and non-metallic minerals.**
2. **Write uses of minerals.**

2.4. ENERGY RESOURCES

Learning Objectives

- Give classification of energy resources
- Explain the distribution and production trends of major energy resources
- Focus on strategic oil reserves
- Describe non-conventional energy resources

There are various primary energy sources, including nuclear energy, renewable sources like wind, solar, geothermal, and hydropower, as well as fossil fuels like oil, coal, and natural gas. These sources are converted into electricity, a secondary energy source that powers homes and businesses, through power lines and other transmission infrastructure.

2.4.1. Classification of Energy Resources

Sources of energy can be classified into:

- Renewable Sources;
- Non-renewable Sources.

Nature provides an abundance of renewable energy sources that are also sustainable. These energy sources are environmentally safe and naturally replenishable. Tidal, hydroelectric, wind, solar, geothermal, biomass, and bioenergy are a few examples of renewable energy sources. Any natural resource found beneath the surface of the earth is considered a non-renewable resource. These kinds of energy resources deplete more slowly than they are used. It takes millions of years for them to recover. Coal, oil, and natural gas are the primary instances of non-renewable resources. Hydrocarbon gas liquids, coal, natural gas, petroleum, and nuclear energy are a few examples of non-renewable energy sources.

Example

Even though fossil fuels are easy to find, burning them releases greenhouse gases that contribute to global warming. Nuclear energy is a highly effective energy source, but it produces radioactive waste that can be difficult to dispose of safely. Renewable energy sources like solar and wind power, while intermittent and sometimes requiring storage systems to balance supply and demand, are sustainable. Our homes, companies, and industries rely on energy resources, so it's important to consider the pros and cons of each type of resource.

2.4.2. Distribution and Production Trend of Major Energy Resources

Currently, a wide range of energy sources includes coal, oil, gas, nuclear, hydropower, solar, wind, and biofuels when we think of energy sources. However, a varied energy system is a relatively new development. It is evident if we go back a few centuries that we were dependent on just one or two major energy sources.

The most common form of energy used worldwide up until the middle of the 19th century was traditional biomass, which is the burning of solid fuels like wood, agricultural waste, or charcoal. Coal saw an astounding increase in popularity during the Industrial Revolution. By the turn of the 20th century, coal accounted for about half of global energy production, with biomass accounting for the remaining half.

Global adoption of a wider variety of sources occurred during the 1900s. Hydropower comes after gas and oil. Nuclear energy wasn't included in the mix until the 1960s. The renewable energy sources commonly known as "modern renewables" – wind and solar – were added much later, in the 1980s. This 200-year record of worldwide energy consumption highlights the fact that energy transitions have historically been extremely sluggish. A specific energy source hasn't become the dominant one for many decades, if not a century. Though it was true in the past, there are indications that this is starting to change. A few recent energy shifts occurred swiftly. In 1990, coal power accounted for nearly two-thirds of the electricity generated in the United Kingdom. This had decreased to just under one-third by 2010. And in the ensuing ten years, it dropped to about 1%.

2.4.3. Strategic Oil Reserves

When a natural disaster, accident, or set of economic sanctions strikes, a nation with strategically located petroleum reserves can withstand an abrupt and transient disruption in supply. Stockpiles of extracted crude oil that can be quickly refined into fuels are known as strategic reserves. They are not to be confused with proven oil reserves, which are estimates of the amount of crude that may eventually be extracted from the earth. Strategic reserves usually aren't able to completely compensate for disruptions that last longer than a few months because storage facilities are constrained in comparison to petroleum consumption. Instead of a continuous increase in supply, releases from reserves signify a one-time addition to it.

Remember

Energy can be renewable but not necessarily sustainable at the same time, e.g. river water saturated with biodiversity markers when combined with hydro-power dam systems along its course, non-regenerative tree plantations, etc.

Did you know?

The world's population is projected to grow to 9.7 billion by 2050, which will lead to increased demand for energy. It is essential that we develop a mix of energy sources to meet the world's growing needs.

2.4.3.1. Examples of Strategic Petroleum Reserves

Did you know?

The U.S. SPR is stored in naturally-occurring salt caverns found deep underground rather than traditional tanks. Salt formations offer the lowest cost, most environmentally secure way to store crude oil for long periods.

In the US, SPR is spread across four enormous salt dome caverns on the Gulf Coast. Founded in 1973 to address the oil crisis brought on by the Arab Oil Embargo, the SPR didn't get its first shipment of crude until 1977. Following Hurricanes Katrina in 2005 and Gustav in 2008, which interrupted energy production in the Gulf Coast region, the government ordered releases from the SPR. Additionally, the U.S. took funds from the SPR in 2022 in reaction to increased gasoline costs and inflation as a result of the disruptions brought on by Russia's invasion of Ukraine.

China is believed to possess the second-largest strategic reserve in the world, even though it does not consistently release information about its size or holdings. China reportedly had enough crude in its strategic reserves by the end of 2021—roughly 490 million barrels—to cover 40 or 50 days' worth of imports.

The International Energy Agency (IEA) estimates that as of the end of 2021, Japan's public strategic reserves were equal to 133 days' worth of crude oil imports. Based on the nation's average monthly imports in 2021, that translates to 332 million barrels. It is interesting that 86 days' worth of imports were covered by commercial stocks.

The IEA mandates that its member nations maintain strategic reserves equivalent to ninety days' net imports from the previous year. It is possible to include commercial stockpiles in this amount. As of year-end 2021, net importers within the organization possessed strategic reserves equivalent to 156 days of imports, including 82 days of commercial inventories.

2.4.3.2. Releasing Reserves From the SPR

The Energy Policy and Conservation Act permits a U.S. president to order a full drawdown of the reserve or a limited one of up to 30 million barrels in the event of "a severe energy supply disruption" to prevent such disruptions. The Department of Energy (DOE) may carry out test drawdowns of up to 5 million barrels under the same statute. Loans or direct sales of oil are two ways in which the SPR releases oil. In a loan or exchange, a company, usually a refiner, receives crude from the reserve; in return, the refiner is required to return the same amount of crude plus extra barrels as interest. Exchanges usually occur when there is a restricted disturbance, like a blockage in a shipping channel, and they must be considered beneficial to the general public.

Additionally, through an online competitive bidding process, the DOE may release crude oil by selling it directly to commercial suppliers. Following a presidential proclamation directing an emergency sale, the Department of Energy has thirteen days to hold the sale, choose bidders, grant contracts, and arrange delivery.

2.4.3.3. Replenishing Strategic Petroleum Reserves

The Department of Energy must defend national security, support domestic supply, keep costs to a minimum, and avoid upsetting markets or market prices when acquiring crude for the SPR. With the COVID-19 pandemic driving down demand and oil prices, then-President Donald Trump directed the Department of Energy to buy 77 million barrels of crude oil to fill the SPR to capacity in March 2020. However, the acquisition plans were abandoned since Congress never supplied the required money. This was regrettable since, in 2022, as a result of Russia's invasion of Ukraine and rising oil prices, the SPR was required.

2.4.4. Non-Conventional Energy Resources

Renewable energy sources are another name for non-conventional energy sources. Tidal energy, wind energy, solar energy, and bioenergy are a few examples of unconventional energy sources.

2.4.4.1. Solar Energy

Sunlight generates solar energy. The amount of sunlight that the photovoltaic cells receive depends on the type of electricity that needs to be generated. The energy is used for water distillation and cooking.

2.4.4.2. Wind Energy

To operate water pumps for irrigation, wind energy is produced by harnessing the power of the wind. In terms of wind power generation, India is currently the second-largest country.

2.4.4.3. Tidal Energy

Utilizing the sea's tidal waves to one's advantage produces tidal energy. Due to a lack of affordable technology, this source isn't being used.

Question

What is the importance of the non-conventional source of energy?

Answer

Non-conventional sources of energy are considered to be important as they are renewable, pollution-free, abundantly available, and environmentally friendly.

Did you got it?

1. **What sources of energy are renewable?**
2. **Can any source of energy be pollution-free?**
3. **Which is the largest non-conventional source of energy?**

2.5. NUCLEAR ENERGY RESOURCES

Learning Objectives

- Understand nuclear fission
- Discuss nuclear waste
- Know the role of the IAEA

Nuclear energy is the kind of energy that is significantly released during processes involving atomic nuclei, or the dense centers of individual atoms. It differs from other atomic phenomena' energy, which only includes the atoms' orbital electrons, like typical chemical reactions. One way to release nuclear energy is through carefully regulated nuclear fission in reactors, which are currently used to produce electricity in many regions of the world. By 2020, controlled nuclear fusion—an additional means of producing nuclear energy—still has room for improvement, but it shows promise. Fusion and fission of nuclear materials have both produced massive releases of nuclear energy.

In nuclear fission, the nucleus of an atom, such as that of uranium or plutonium, breaks up into two lighter nuclei of roughly equal mass. The process may take place spontaneously in some cases or may be induced by the excitation of the nucleus with a variety of particles (e.g., neutrons, protons, deuterons, or alpha particles) or with electromagnetic radiation in the form of gamma rays. In the fission process, a large quantity of energy is released, radioactive products are formed, and several neutrons are emitted. These neutrons can induce fission in a nearby nucleus of fissionable material and release more neutrons that can repeat the sequence, causing a chain reaction in which a large number of nuclei undergo fission and an enormous amount of energy is released. If controlled in a nuclear reactor, such a chain reaction can provide power for society's benefit. If uncontrolled, as in the case of the so-called atomic bomb, it can lead to an explosion of awesome destructive force.

The process by which lighter elements undergo nuclear reactions to form heavier elements is known as nuclear fusion. In situations where the interacting nuclei are from low-atomic-number elements (e.g., significant amounts of energy are released, such as in the case of hydrogen [atomic number 1] or its isotopes tritium and deuterium). Thermonuclear weapons, also known as hydrogen bombs, were the first to use the enormous energy potential of nuclear fusion. They were created in the years immediately following World War II. Nuclear fusion has enormous potential for peaceful uses, especially considering the nearly infinite supply of fusion fuel on Earth. This has led to a great deal of effort being put into harnessing the process of power generation. The conditions for plasma temperature and heat insulation have mostly been met, even though practical fusion reactors have not yet been constructed. This suggests that the possibility of using fusion energy to produce electricity is now very real. Commercial fusion reactors hold out the prospect of providing nations with an endless supply of electricity.

The Role of the IAEA:

- The International Atomic Energy Agency (IAEA) establishes and promotes international standards and guidance for the safe and secure use of nuclear energy to protect people and the environment.
- The IAEA supports existing and new nuclear programs around the world by providing technical support and knowledge management. Through the Milestones Approach, the IAEA provides technical expertise and guidance to countries that want to develop a nuclear power program as well as to those who are decommissioning theirs.
- Through its safeguards and verification activities, the IAEA oversees that nuclear material and technologies are not diverted from peaceful use.
- Review missions and advisory services led by the IAEA provide guidance on the activities necessary during the lifetime of production of nuclear energy: from the mining of uranium to the construction, maintenance, and decommissioning of nuclear power plants and the management of nuclear waste.
- The IAEA administers a reserve of low-enriched uranium (LEU) in Kazakhstan, which can be used as a last resort by countries that are in urgent need of LEU for peaceful purposes.

Keyword

Nuclear fusion is the process by which two light atomic nuclei combine to form a single heavier one while releasing massive amounts of energy.

Question

What is nuclear fission?

Answer

Nuclear fission is a reaction where the nucleus of an atom splits into two or more smaller nuclei while releasing energy.

Did you get it?

1. **What are the advantages of nuclear energy?**
2. **Why are nuclear power plants considered among the safest and most secure facilities in the world?**

ROLE MODEL

ENRICO FERMI

Italian-American physicist

On December 2, 1942, Enrico Fermi and his team of scientists harnessed the atom and opened the door to new scientific and technological realms. His achievement allowed the U.S. to produce the atomic bomb which helped end World War II. Now, more than fifty years later, nuclear energy provides a significant part of the world's electrical power, and radioactive materials are used for hundreds of industrial, agricultural, and medical applications—from food preservation to cancer therapy, checking the integrity of welds in pipelines and bridge supports, and gauges that measure the thickness of coatings applied to paper.

Fermi was born in Rome in 1901 and received his doctorate from the University of Pisa in 1922. From 1923 to 1924, he studied in Göttingen, Germany, with Max Born, and in Leiden, The Netherlands, with Paul Ehrenfest. From 1924 to 1926, Fermi lectured in mathematical physics and mechanics at the University of Florence. He became the first professor of theoretical physics at the University of Rome, where he taught for 12 years.

In 1938, Fermi traveled to Stockholm to receive the Nobel Prize "for his identification of new radioactive elements produced by neutron bombardment and his discovery, made in connection with this work, of nuclear reactions effected by slow neutrons." Fermi's wife Laura, in her book *Atoms in the Fam*ily*, wrote, "Four years of patient researches; the broken and the unbroken tubes full of beryllium powder and radon; the strenuous races along the hall of the physics building to rush element after element to the Geiger counters; the efforts to understand nuclear processes, and the many tests to prove the theories ... these had won the Nobel Prize for Enrico."*

By 1938, Fascist Italy had become intolerable to Fermi. He used his trip to Stockholm as an opportunity to flee. He and his family sailed directly to the United States, where he had visited many times before. In the months before his Stockholm trip, he had sought teaching positions in America and had received offers from five universities. He chose a professorship in Physics at Columbia University. There, in 1939, he confirmed the discovery of the fission process and began striving to attain a nuclear chain reaction.

From 1942 to 1944, Fermi worked at the Metallurgical Laboratory of the University of Chicago. In a makeshift laboratory under Stagg Field Stadium, he designed and built the first nuclear reactor and led the epochal experiment that demonstrated the first self-sustained chain reaction. More than any individual, he was responsible for developing a means for the controlled release of nuclear energy.

Fermi became a U.S. citizen in 1944. From 1944 to 1945, he became Associate Director of the Los Alamos National Laboratory in New Mexico. In 1946, he returned to the University of Chicago as a professor at the Institute of Nuclear Studies, which now bears his name--Fermi Institute. He resumed his fundamental research interests in nuclear and elementary particle physics and also, beginning in 1950, served as one of the first members of the General Advisory Committee of the Atomic Energy Commission.

On November 16, 1954, President Eisenhower and the Atomic Energy Commission gave Fermi a special award for his lifetime of accomplishments in physics and, in particular, for the development of atomic energy. Fermi's other research resulted in the Fermi-Dirac particle statistics, the theory of beta-decay, the Thomas-Fermi model of the atom, and a theory of the origin of cosmic rays.

Fermi died on November 28, 1954, and the Enrico Fermi Award was established in 1956 to perpetuate the memory of his brilliance as a scientist and to recognize others of his kind-inspiring others by his example.

FOCUS ON CAREERS

WATER RESOURCES ENGINEER

For water resource management facilities across the country, water resource engineers create new machinery and systems. Water resource engineers design systems that guarantee the public always has access to clean, uncontaminated water for drinking, housing, and recreational uses. In addition to designing these water management systems, water resource engineers frequently supervise their installation and upkeep. This rapidly expanding industry is driven by the ongoing need for more water due to population growth. Although many Water Resource Engineers also go on to pursue their Masters' Degrees, a Bachelor's degree and official certification are prerequisites for this career.

Role of Water Resource Engineer

Within the specialized field of civil engineering known as "water resource engineering," new apparatus and systems for the management of human water resources are designed. Water resource engineers work in a variety of fields, including natural springs, underground wells, and water treatment facilities. To improve water treatment and aquatic resource management's efficacy and efficiency, water resource engineers must develop new machinery and procedures. During a regular workday, data from pertinent areas is analyzed, and new or improved facilities are designed to maximize the water treatment system's cleaning capabilities. When designing these systems, a water resource engineer must take government regulations, budgetary restrictions, and other factors into account. The development and installation of these systems may then be supervised by a water resource engineer to guarantee correct assembly. Once finished, they might oversee these systems' upkeep.

Where Does a Water Resource Engineer Work?

The majority of a water resource engineer's workday is spent in an office reviewing data and creating new water resource management systems. To enable the engineer to supervise the implementation of their designs, a portion of the workday might be dedicated to construction sites. Additionally, when overseeing maintenance on cutting-edge machinery, they might find themselves in more industrial settings. Some water resource engineers decide to take part in major engineering projects overseas. The majority of water resource engineers are full-time employees, and many of them work over 40 hours per week. They can effectively oversee projects and ensure that everything is going according to plan thanks to this additional time.

Average Salary of Water Resource Engineer

As of May 2020, water resource engineers, who are classified as environmental engineers by the BLS, made a median salary of $92,120. Over this period, the top 10% earned more than $144,670, while the bottom 10% made roughly $55,450.

Job Demand for Water Resource Engineers

Between 2020 and 2030, there is expected to be a 4% increase in the demand for water resource engineers' jobs. Because infrastructure ages over time, it must be rebuilt and repaired. Furthermore, as our population grows and our water resources are continually strained, we will need more Water Resource Engineers to design new, more effective systems.

Education Requirements to Become a Water Resource Engineer

At the very least, a bachelor's degree is required of water resource engineers. To aid in the licensure process—which is mandatory for all engineers in some states—you may also be required to enroll in an ABET-accredited program.

Typically, becoming licensed requires passing the Fundamentals of Engineering (FE) Exam, followed by a specified period spent as an Engineer-in-Training (EIT) or Civil Engineering (CE) Intern. Each state has different requirements for obtaining a license. A large number of water resource engineers continue to earn master's degrees. Actually, one out of every five civil engineers goes to college. This raises your chances of being hired as well as your chances of getting hired for a management position that pays more.

A CLOSER LOOK

WATER SCARCITY

Water scarcity is a common occurrence, even in nations with sufficient water resources. It is obvious that climate change, along with human factors, is increasingly depriving children of their right to safe water and sanitation, even though there are many possible causes for this, including failing infrastructure and distribution systems, contamination, conflict, or inadequate management of water resources. Water scarcity restricts access to clean water for drinking and basic hygiene practices at home, in schools, and in medical facilities. When there is a shortage of water, cholera outbreaks can occur and sewage systems can malfunction. Additionally, the cost of scarce water increases. Women and children suffer more from water scarcity because they are frequently the ones who have to gather it. Water that is farther away takes longer to get, which frequently translates into less time spent in class. Lack of water in schools affects student enrollment, attendance, and performance, especially for girls. Long-distance water carrying is extremely taxing physically and puts kids in danger of being exploited.

UNICEF addresses the complexities and differences in water scarcity drivers by implementing context-specific technologies and initiatives to improve access to safe water and mitigate the effects of water scarcity at various levels within countries and regions. UNICEF focuses on:

- Identifying new water resources: Using a variety of technologies, such as remote sensing, geophysical surveys, and field research, UNICEF evaluates the availability of water resources.
- Improving the efficiency of water resources: In order to minimize water leakage and contamination, UNICEF repairs urban water distribution networks and treatment systems. It also encourages the reuse of wastewater for agricultural purposes, thereby safeguarding groundwater.
- Planning for urban scarcity: UNICEF reduces the risk of cities running out of water by identifying available resources and planning for future water needs.
- Expanding technologies to ensure climate resilience: UNICEF works to create and support climate-resilient water sources, such as solar-powered water networks that tap into deeper groundwater reserves. UNICEF also advances water storage through rainwater harvesting, managed aquifer recharge, and small-scale retention structures. Managed aquifer recharge involves pumping water into underground reserves to improve its quality.
- Changing behaviors: Through supporting environmental clubs in schools, among other initiatives, UNICEF works with communities and schools to raise awareness of the value of water and the necessity of protecting it.
- Planning national water needs: UNICEF advocates for ensuring that national planning considerations take into account the water requirements for domestic use, health, and sanitation by working with key stakeholders at the national and sub-national levels.

SUMMARY

- Land resources refer to those that come directly from the land itself. These encompass raw materials like coal, bauxite, gold, and other minerals; underground water; and agricultural land that naturally enriches the crops grown.
- The land use pattern is shaped by various factors such as climate, soil, population density, and technical and socioeconomic characteristics. These factors influence how the land is utilized and organized.
- Land, like air and water, is crucial for humanity. Land degradation, the decline in soil productivity, leads to food insecurity, higher food prices, climate change, environmental threats, loss of biodiversity, and impacts on ecosystem services.
- Land plays a crucial role for those living in rural and urban areas who are struggling financially, serving as a key foundation for their social and economic progress. Fair access to land and secure property rights can also help empower these individuals in facing the challenges brought about by globalization.
- Water is crucial for all living beings. While it's a renewable resource, there are still many regions facing a serious lack of clean water.
- Water resources are facing significant pressure in most countries today. The global population is growing rapidly, and estimates suggest that by 2030, there will be a 40% disparity between the world's projected water demand and supply due to current practices.

A scarcity of potable water for a population is termed a water crisis. Drought, famine, and loss of life are the eventual consequences of this situation. In drought-affected regions and on the African continent, access to clean drinking water is now seen as a luxury.

- Water is vital for human survival.. Although freshwater is abundant, its supply is limited, and as a result of the rising global population and increasing economic levels, freshwater demand has been rising rapidly.
- Water is essential for our daily lives. There's always space for conservation when it comes to using water, whether it's inside or outside your home or workplace. The fundamental foundation for socioeconomic progress lies in mineral resources. When discussing mineral resources, we typically categorize them into metallic resources and non-metallic resources.
- Mineral deposits determine their usage. Some nations lack them, while others have abundant deposits.
- Each mineral typically forms fully developed crystals and is treated based on its crystal type.
- Primary energy sources include nuclear, renewable (wind, solar, geothermal, hydropower), and fossil fuels (oil, coal, natural gas).
- Energy sources are categorized as renewable or non-renewable.
- Currently, we have a wide range of energy sources like coal, oil, gas, nuclear, hydropower, solar, wind, and biofuels to consider. However, having a diverse

energy system is a more recent advancement. In the event of a natural disaster, accident, or economic sanctions, a country with strategically located petroleum reserves can endure a sudden and temporary supply disruption.

- Non-conventional energy sources are also known as renewable energy sources. Tidal energy, wind energy, solar energy, and bioenergy are a few examples of unconventional energy sources.
- Nuclear energy is the kind of energy that is significantly released during processes involving atomic nuclei, or the dense centers of individual atoms.

MULTIPLE CHOICE QUESTIONS

1. **On the basis of origin, resources can be classified as ______ and _____.**
   a. Biotic and Abiotic
   b. Renewable and Non-renewable
   c. Potential and Developed
   d. None of the above
2. **On the basis of the status of development, resources can be classified as ______.**
   a. Potential
   b. Developed Stock
   c. Reserves
   d. All of the above
3. **Which of the following statements is true?**
   a. Abiotic resources are obtained from the biosphere and have no life, such as rocks and metals.
   b. Biotic resources are all those things which are composed of living things like human beings, flora, and fauna.
   c. Both options (a) and (b) are false.
   d. Both options (a) and (b) are true.
4. **Minerals and fossil fuels are examples of non-renewable resources which take _____ for their formation.**
   a. Hundreds of years
   b. Thousands of years
   c. Millions of years
   d. Just a year
5. **All the minerals, water resources, forests, wildlife, land within the political boundaries, and oceanic area up to __________ from the coast are termed as territorial waters, and the resources therein belong to the nation.**
   a. 20 nautical miles
   b. 25 nautical miles

c. 15 nautical miles

d. 12 nautical miles

6. The oceanic resources beyond ______________ of the Exclusive Economic Zone belong to the open ocean, and no individual country can utilize these without the concurrence of international institutions.

a. 200 nautical miles

b. 12 nautical miles

c. 22.5 nautical miles

d. 100 nautical miles

7. ______ are the resources that are surveyed, and their quality and quantity have been determined for utilization.

a. Developed resources

b. Reserves

c. Stock

d. Abiotic

8. _______ are the subset of the ______, which can be put into use with the help of existing technical 'know-how,' but their use has not been started.

a. Reserves, developed resources

b. Stocks, reserves

c. Developed resources, stock

d. Reserves, stock

REVIEW QUESTIONS

1. What is sustainable development?
2. How can you divide resources?
3. What are the patterns of land use?
4. How is land use pattern changed?
5. What is the land use change method?
6. What is the state of water resources?
7. What are the main types of mineral resources?
8. What is the importance of mineral resources?
9. What are the two types of energy resources?
10. How are energy resources used?

Answer to Multiple Choice Questions

1. (a) 2. (d) 3. (c) 4. (c) 5. (d)

6. (a) 7. (a) 8. (d)

REFERENCES

1. Bauman, A., (2008). Natural Resources. Tunbridge Wells: Ticktock.
2. Craig, J., Vaughan, D., & Skinner, B., (2011). Earth Resources and the Environment (4th edn., p. 92). Pearson Prentice Hall.
3. Gillett, J., (2013). Natural Resources. New York: PowerKids Press.
4. Jakab, C., (2007). Natural Resources. North Mankato, MN: Smart Apple Media.
5. Sally, M., (2009). Natural Resources. Mankato, Minn: Black Rabbit Books.
6. Sasan, S., & Sara, G., (2021). Reclaiming Hydrogen for a Renewable Future.
7. Stille, D. R., (2007). Natural Resources. Mankato: Compass Point Books.

CHAPTER

3 Agriculture

LEARNING OBJECTIVES

After studying this chapter, you will be able to:

- Understand the benefits and role of agriculture
- Identify the location of agriculture
- Explain world agriculture typologies
- Focus on factors affecting agriculture
- Discuss food security

"The greatest advances of civilization, whether in architecture or painting, in science and literature, in industry or agriculture, have never come from centralized government."

—Milton Friedman

INTRODUCTORY EXAMPLE

Origins of Agriculture

Agriculture is the active production of useful plants or animals in ecosystems that humans have created. A more holistic viewpoint maintains that humans are environmental engineers who disrupt terrestrial habitats in specific ways. Agriculture has historically been conceptualized narrowly, in terms of particular combinations of activities and organisms—wet-rice production in Asia, wheat farming in Europe, cattle ranching in the Americas, etc. Different localized changes are brought about by anthropogenic disturbances like clearing vegetation or tilling the soil; typical outcomes include a rise in the amount of light that reaches the ground and a decrease in interspecies competition. Because of this, a region might produce more of the plants or animals that people need for technology, medicine, food, and other purposes.

Certain plants and animals have evolved into domesticated species, meaning they now rely on human actions and other resources for long-term survival. The biological process known as "domestication" occurs when organisms acquire traits through human selection that make them more useful, such as plants producing bigger seeds, fruits, or tubers than their wild counterparts. The term "cultigens" refers to domesticated plants, which can be found in many different families (groups of closely related genera that share a common ancestor). Because of their traits that make them especially conducive to domestication, the families of grasses (Poaceae), beans (Fabaceae), and nightshade or potato (Solanaceae) have produced an abnormally high number of cultigens.

Generally speaking, domesticated animals descended from social species in the wild that could be bred to increase desirable traits for humans, just like plants. In addition to being more tame than their wild counterparts, domesticated animals also frequently yield more milk, meat, or wool. In addition to being a symbol of wealth, they have been used for traction, transportation, pest control, help, and companionship. Several species have a large number of domesticated variations, or breeds: Canis lupus familiaris (dog), Felis catus (cat), Bos species (cattle), Ovis species (sheep), Capra species (goat), Sus species (swine), Equus caballus (horse), Gallus gallus (chicken), and Anser (duck and goose).

Agriculture has changed significantly over time and space because it is a cultural phenomenon. Domesticated plants and animals have been raised at a variety of scales, from small-scale domestic operations to large-scale commercial endeavors, and they still are.

INTRODUCTION

Commonly known as farming, agriculture is a thoughtful effort to reshape a part of Earth's surface by growing plants and crops, as well as raising livestock for economic purposes and human needs. Some species of ants, termites, and beetles have been cultivating crops for 60 million years before humans, as commonly understood. Using natural resources to "create goods that support life, such as food, fiber, wood products, crops, and related services" is how it's described, though with different focuses. This includes gardening, animal care, forestry, and crop cultivation.

3.1. BENEFITS AND ROLE OF AGRICULTURE

Learning Objectives

- Know why is agriculture important?
- Discuss how is agriculture important
- Know the importance of agriculture in everyday life
- Explain how agriculture affects the economy.
- Understand the importance of agricultural biodiversity

Cultivating natural resources for human sustenance and financial benefit is known as agriculture. It blends new technologies and contemporary production techniques with the ingenuity, imagination, and skill required to raise animals and plant crops. Commodities, or basic goods used in trade, like grain, livestock, dairy, fiber, and raw materials for fuel, are another service that agriculture offers to the world economy. For instance, fiber is a major crop, which is an essential resource for the apparel industry in agricultural production in the U.S., as stated by The Balance Small Business.

3.1.1. Why Is Agriculture Important?

The output of agriculture, which includes raw material production, participation in the global supply chain, and economic development, is a major factor in the industry's and society's importance.

3.1.1.1. Providing Raw Materials

The basic elements of the world economy are raw materials. Manufacturers need access to raw materials to produce goods. Steel, minerals, and coal are examples of non-agricultural raw materials. However, agriculture also provides many raw materials, such as lumber for construction and herbs for flavoring food. For example, corn is used for food and ethanol production. Another example is plant-based *resins* used in various industrial applications like paints, coatings, and adhesives for construction.

3.1.1.2. Creating a Strong Supply Chain

Shipping methods like rail, trucking, and ocean freight are needed for the import and export of goods such as agricultural products. The global supply chain may be impacted by delays in the shipment of agricultural products from a port in Los Angeles to China and vice versa. For example, according to the Iowa Soybean Association, delays in the shipments of crops from South America contributed to the significant increase in sales of soybean crops in Iowa in 2021. Iowa gained a competitive advantage in this instance. However, delays in crop shipping could also harm areas awaiting shipments, reducing the availability of goods on store shelves and negatively affecting livelihoods.

Keyword

Resin is a solid or highly viscous substance of plant or synthetic origin that is typically convertible into polymers.

3.1.1.3. Encouraging Economic Development

The connection between agriculture and other economic sectors, which promotes job creation and economic development, has an impact on global trade. USAID states that nations with robust agricultural sectors see increases in employment across various industries. Since farmers in these nations innovate through technology and farm management techniques to boost agricultural productivity and profitability, countries with growing agricultural productivity and strong agricultural infrastructure also have higher per capita incomes.

3.1.1.4. Resources on the Importance of Agriculture

- American Farm Bureau Federation, Fast Facts About Agriculture & Food: Provides various statistics demonstrating why agriculture is important.
- The Western Producer, "Suddenly Agriculture Is Important": Highlights agriculture's role as a stable commodity provider even amid disruption.
- LinkedIn, "What Is Agriculture and Its Importance?": Discusses the importance of agriculture in 10 areas.

3.1.2. How Is Agriculture Important?

The tech sector gets a lot of focus when disruptions happen in global supply chains. For example, a manufacturer's production of computers, cars, and other items is affected by the shortage of computer chips, which are made from silicon, a non-agricultural raw material. This impacts various aspects of business and society.

In a global economy where economies are interconnected, agriculture is also essential to meeting consumer and business market demand. These are a few different kinds of agricultural products.

3.1.2.1. Fruits and Vegetables

Fiber, proteins, and carbohydrates that are necessary for human diets can only be found in fruits and vegetables. Many fruits and vegetables naturally contain vitamins (like A, C, and E) and minerals (like magnesium, zinc, and phosphorus). Fruits and vegetables not only improve health but also broaden the palate's flavor profile.

Keyword

A **legume** refers to any plant from the Fabaceae family that would include its leaves, stems, and pods.

3.1.2.2. Animal Feed

Some fruits and vegetables are farmed for animal feed, such as poultry and cattle. According to the American Feed Industry Association, approximately 900 ingredients for animal feed have legal approval in the U.S. These consist of agriculturally produced items like oils, *legumes*, sprouted grains, hay, and straw.

3.1.2.3. Natural Rubber Production

Based on market research conducted by Hedges and Company, there are over 1.4 billion automobiles worldwide. All of them have rubber tires on their wheels. As per GEP, small-scale farmers account for approximately 90% of suppliers, and the top three countries producing rubber are Thailand, Indonesia, and Malaysia, which together account for around 70% of the world's natural rubber production.

3.1.2.4. Cotton for Clothing

Production in agriculture is the first step on the journey from cotton to clothing. Before becoming a piece of clothing, cotton must be cultivated, harvested, processed, spun, and woven into fabric. A vast global supply chain supports the production of cotton, which is a major commodity that accounts for about 31% of all textile fibers worldwide, according to Forum for the Future.

3.1.2.5. Biofuels

The Environmental Protection Agency (EPA) reports that biofuels have favorable economics when made from biomass sources, which include agricultural products like algae, corn, soybeans, and sugarcane. The advantages include the possibility of higher farmer incomes as well as a decrease in greenhouse gas and pollutant emissions. However, the use of water and land resources in the production of biodiesel may increase the cost of food.

3.1.2.6. Industrial Products

Utilizing biomass as a source of raw materials to create industrial products is known as "bio-based chemistry." A variety of industrial goods made from bio-based chemicals include fertilizers, inks, dyes, plant oils, biolubricants, and bioplastics. Chemicals and goods made from biomass provide an alternative to traditional goods made from petroleum. Because it encourages the minimization of environmental effects in industrial production, bio-based chemistry is regarded as a form of green chemistry.

3.1.2.7. Pharmaceutical Products

For thousands of years, people have used plants as medicine to help heal their ailments. For instance, the plant root ginger, which is usually ingested in tea, can facilitate better digestion. Herbal and plant-based products have additional medicinal uses. For instance, the heart failure medication digoxin is made from chemicals that have been extracted from the foxglove plant. Another illustration is the chemical known as polylactic acid (PLA), which is created in green plants when glucose is fermented into lactic acid. According to a study published in Engineered Regeneration, PLA has uses in orthopedic interventions, cancer therapy, cardiovascular implants, *tissue engineering*, and surgical implant production.

3.1.3. Importance of Agriculture in Everyday Life

It is impossible to downplay the significance of agriculture in daily life. Things like cleaning the house and getting ready for work would not be possible without the agriculture industry. Examples of agricultural products we use daily are as follows:

- Shelter. Wood and plant-based materials, such as bamboo, can be used for indoor décor and construction materials.
- Morning routine. Mint is often an ingredient in toothpaste, adding flavor while brushing your teeth, and the caffeine in coffee that keeps you awake is derived from the coffee bean.
- Dressing up. In addition to cotton, clothing can be manufactured from hemp, ramie, and flax. Bio-based materials can be used to produce grooming products such as skin creams and shampoos.

Keyword

Tissue engineering is a biomedical engineering discipline that uses a combination of cells, engineering, materials, methods, and suitable biochemical and physicochemical factors to restore, maintain, improve, or replace different types of biological tissues.

- Cleaning. Two types of chemicals used in detergents, cleaning products, and bath or hand soap—surfactants and solvents—can be produced from biomass.
- Driving to work. Plants make it possible to get to and from work. Think of rubber (sourced from rubber trees) and biodiesel fuel, which often includes ethanol (sourced from corn).
- Entertainment. Paper from trees enables you to write, and some musical instruments, such as reed instruments, require materials made from plants.
- Education. From pencils (still often made of wood) to paper textbooks, students rely on agricultural products every day.

Keyword

Biodiversity is all the different kinds of life you'll find in one area—the variety of animals, plants, fungi, and even microorganisms like bacteria that make up our natural world.

3.1.4. How Does Agriculture Affect the Economy?

The economy can be significantly impacted by agriculture. According to the Department of Agriculture's (USDA) Economic Research Service, the food and agriculture industries accounted for 10% of all employment in the US, estimated at 20 million full- and part-time positions in 2020. Furthermore, according to USDA data, crop revenue in 2020 was close to $198 billion. Revenues from animals and animal products came in close behind in 2020, coming in at $165 billion.

The Cybersecurity and Infrastructure Security Agency (CISA) states that the food and agriculture sector is an essential driver of economic activity due to its interdependence with other industries, such as energy, chemicals, transportation, water and wastewater systems, and water. Agriculture plays a significant role in the economic development of the United States, making direct and indirect contributions to the gross domestic product (GDP). It achieves this through multiple sectors including farm production, forestry, fishing activities, textile mills and products, apparel, and sales of food and beverages. Additionally, agriculture also supports service and manufacturing industries.

- Farm production. According to the most recent USDA data on farming and farming income, the U.S. had 897 million acres, or just over 2 million farms, in 2020. To meet the demand for agricultural products both domestically and internationally, farms produce a variety of fruits, vegetables, plants, and crops.
- Forestry and fishing activities. Forestry and fish harvesting in water farms or in the fish's natural habitat are examples of agricultural activities. As per the USDA, agroforestry aims to accomplish goals, needs, and values by "establishing,

managing, using, and conserving forests, trees, and associated resources in a sustainable manner." Aquaculture is a type of fishing where food is produced by raising fish and other marine life in controlled environments.

- Textile mills and products. The U.S. cotton industry produces $21 billion in products and services annually, according to the USDA. The industry has created various employment roles, such as growers, ginners, and buyers working on farms and in textile mills, cotton gins, offices, and warehouses.
- Apparel and food and beverage sales. Sales of goods derived from agricultural production are crucial because agriculture is a business. Food services and eating and drinking establishments accounted for 10 million jobs in 2020, the largest share of all categories within the agriculture and food sectors, according to the USDA. As such, assisting growers in building capacity and understanding market dynamics is a critical component of the sales component in agriculture. Many of these customers have high expectations.
- Manufacturing. A vast array of products, such as textiles, building materials, fuels, cleaning and personal care items, food and beverage products, and more, are made using agricultural products. The USDA estimates that approximately 11.7 million people are employed by food and beverage manufacturing companies in the U.S.

Here are ways agriculture and related industries impact economic development.

3.1.4.1. Agribusiness

Businesses engaged in the commercial tasks necessary to get agricultural products to market are referred to as agribusiness. It encompasses all kinds of food-related enterprises, from local family farms to large international conglomerates. Within the U.S., according to the USDA, farms made up roughly 0.6% of the total GDP in 2019, or $136 billion.

But farms are only one part of agribusiness. Agribusiness also encompasses companies that produce and refine biofuels, as well as those that manufacture chemical-based goods like fertilizers and agricultural equipment such as tractors. According to USDA data, the GDP contribution of farms and related industries in 2019 was over $1.1 trillion, or slightly more than 5% of the total GDP. Developing supply chains and production networks that support national economic and social stability is another aspect of agribusiness economics. Agribusiness is essential to maintaining *biodiversity* and the environment around farms, as well as to the sustainable use of natural resources, thanks to its advancements in organizational and technological knowledge.

3.1.4.2. Food Security

In order to achieve the Sustainable Development Goals (SDGs), particularly SDG 2: Zero Hunger, sustainable agriculture is essential to the agricultural sector and to ensure food security. The World Bank claims that, in addition to ensuring food security, the agricultural sector increases the incomes of the poorest communities up to four times more successfully than other sectors.

3.1.4.3. Job Creation

Agriculture is a major contributor to job creation globally. For example, a report about the importance of agribusiness from BBVA, a corporate and investment bank, states that agriculture accounts for 25% of exports in developing countries in Latin America or roughly 5% of their regional GDP. In these nations, this activity generates jobs and economic activity.

Did you know?

In the U.S., agriculture and related industries provide 19.7 million full- and part-time jobs, about 10.3% of all employment.

3.1.4.4. Resources on the Economic Impact of Agriculture

The following resources highlight agriculture's impact on the economy, from how disruption affects the business and the benefits of the sector to people's livelihoods:

- Economic Research Service, Farming and Farm Income: Provides an overview of trends in farming and economic development statistics.
- American Journal of Agricultural Economics, "The Importance of Agriculture in the Economy: Impacts from COVID-19": Highlights why agriculture is important based on the impact of COVID-19's disruptions to the sector.
- Canadian Journal of Agricultural Economics, "Agriculture, Transportation, and the COVID-19 Crisis": Discusses how transportation services that COVID-19 has disrupted can impact agricultural supply chains.

3.1.5. Importance of Agricultural Biodiversity

Crop yields have increased as a result of improved farming machinery and increased use of pesticides and fertilizers. Simultaneously, they have caused pollution to soil and water, as well as contributing to climate change. According to NASA, climate change will result in a 24% decrease in corn crop yields by 2030. Maintaining a healthy biodiversity, however, can lessen the effects. Some things to think about are as follows:

- Sustainable agriculture. Farmers and ranchers can increase soil fertility, support good environmental practices, reduce environmental impacts through climate action, and help ensure the profitability of their land by implementing sustainable agricultural practices.
- Climate change regulation. Roughly 10% of U.S. production came from the agricultural sector. The EPA's estimate of greenhouse gas emissions in 2019. In addition to offering direction on agricultural adaptation to meet the challenges

posed by climate change, regulations and policy changes can support the adoption of sustainable practices in the industry.

- Agriculture technology and innovation. From temperature- and moisture-sensing devices to GPS technologies for land surveys to robots, agriculture technology can result in higher crop yields, less chemical runoff, and lower impact on natural resources.

Question

Why is agriculture important for the future?

Answer

Agriculture offers an opportunity to improve the lives of millions of food-insecure people and help countries develop economies that create jobs and raise incomes. Currently, agriculture also impacts future generations. To ensure the long-term success of the global agricultural sector, building a more sustainable economic system aligned with the U.N.'s Sustainable Development Goals is a crucial imperative to help create a more equitable society.

Did you get it?

1. **How is agriculture important?**
2. **How does agriculture affect the economy?**

3.2. LOCATION OF AGRICULTURE

Learning Objectives

- Explain the types of agricultural regions
- Give examples of agricultural regions
- Identify issues in agricultural regions
- Focus on strategies for addressing agricultural regions

The ability to produce enough food to feed a growing population and the amount of land that can be used for agriculture have always required careful balance. What and where crops can be grown is limited by factors like climate, terrain, groundwater, and soil composition unless significant changes to the human landscape are made. The amount of land that is currently under cultivation has greatly increased thanks to new technologies and scientific understanding. The most important factors in determining which land is suitable for particular crops and types of agriculture, however, continue to be spatial variations in land resources like rainfall and temperature zones.

The last 50 years have seen a 12% increase in the amount of land under cultivation worldwide, primarily at the expense of habitats for forests, wetlands, and grasslands. The area that is irrigated has doubled worldwide at the same time. These resources—land and water—are not distributed equally among the nations. While crop production occupies a relatively small portion of the world's land and water, the majority of easily accessible and, therefore, economically valuable resources are either under cultivation or are put to other beneficial uses for both the environment and the economy. As such, the potential to increase the area under cultivation is constrained. The only remaining areas of potential expansion are in sub-Saharan Africa and parts of South America. Additionally, the competition for water resources has been intensifying, to the point where over 40% of rural populations worldwide now reside in areas with limited water supplies.

There are 13.2 billion hectares (ha) of land on Earth. 10,000 square meters is equal to one hectare, which is a common land measurement used in forestry and agriculture in the metric system. Out of this, 12.6 billion hectares (or 12%) are currently used for crop cultivation, 30.7 billion hectares (or 28%) are covered by forests, and 40.6 billion hectares (or 35%) are made up of grasslands and woodland ecosystems. Though they make up only 22% of the world's land area, low-income nations are home to 38% of its people. Land use varies with climatic and soil conditions and human influences.

The distribution of water resources for irrigation is incredibly unequal; some nations have an abundance of water, while others experience severe water scarcity or shortages. Furthermore, even in areas where water might seem plentiful, a large portion of it is either inaccessible, would be very costly to develop, or is not adjacent to land that could be used for farming. There are three types of water scarcity: physical (when the supply cannot meet demand), infrastructural (when existing infrastructure prevents users from meeting their needs), and institutional (when policies and institutions fall short of providing users with a steady, safe, and fair supply of water) (Figure 3.1).

Remember

Water is one of the natural resources, which are found in an adequate amount. It is an essential source for the existence of life on the planet Earth

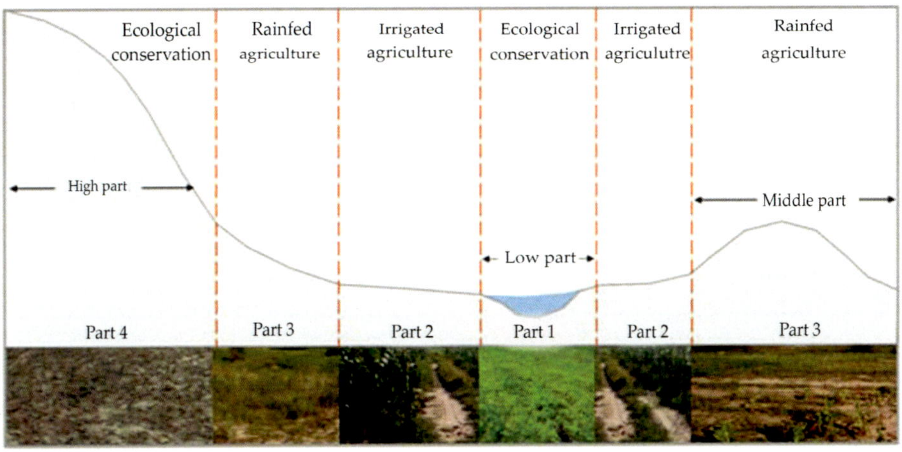

Figure 3.1. Evolution of land under irrigated and rainfed cropping (1961–2008).

Source: By R. Adam Dastrup, MA, GISP, Introduction to Human Geography, https://pressbooks.pub/humangeography/chapter/6-3/ (© 2019 by R. Adam Dastrup, MA, GISP is licensed under a Creative Commons Attribution 4.0 International License, except where otherwise noted).

Countries are already using more water than is available in some regions, especially in the Middle East, Northern Africa, and Central Asia. Ecosystem stresses as a result are becoming more and more obvious. More than 40% of rural people worldwide are thought to reside in river basins where there is physically little water.

Table 3.1. Types of Rainfed Production Systems and Regions

System	Characteristics and Examples
Rain-fed agriculture: **highlands**	Low productivity, small-scale subsistence (low-input) agriculture; a variety of crops on small plots plus a few animals.
Rain-fed agriculture: **dry tropics**	Drought-resistant cereals such as maize, sorghum, and millet are common. Livestock often consist of goats and sheep, especially in the Sudano-Sahelian zone of Africa and in India. Cattle are more widespread in southern Africa and Latin America.
Rain-fed agriculture: **humid tropics**	Mainly root crops, bananas, sugar cane, and notably soybeans in Latin America and Asia. Maize is the most important cereal. Sheep and goats are often raised by more impoverished farmers, while cattle are held by wealthier ones.
Rain-fed agriculture: **subtropics**	Wheat (the essential cereal), fruits (e.g., grapes and citrus), and oil crops (e.g., olives). Cattle are the most dominant livestock. Goats are also essential in the southern Mediterranean, while pigs are dominant in China and sheep in Australia.
Rain-fed agriculture: **temperate**	Principal crops include wheat, maize, barley, rapeseed, sugar beet, and potatoes. In the industrialized countries of Western Europe, the United States, and Canada, this agricultural system is highly productive and often combined with intensive, penned livestock (mainly pigs, chickens, and cattle).

Source: By R. Adam Dastrup, MA, GISP, Introduction to Human Geography, https://pressbooks.pub/humangeography/chapter/6–3/© 2019 by R. Adam Dastrup, MA, GISP is licensed under a Creative Commons Attribution 4.0 International License, except where otherwise noted.

Simultaneously, urban and industrial demand has been growing faster than agricultural demand in more developed nations. While agriculture continues to be the primary use of water in less developed nations, industry uses 55% of the water in Europe. While there are localized water stressors everywhere in the world, some regions—most notably the Middle East, the Indian subcontinent, and northeastern China—face extreme water stress. The Americas and Sub-Saharan Africa suffer from less water stress. The return of runoff to the environment has an effect on the quality of the water as well. Water quality has generally suffered more from rising population and economic growth in conjunction with inadequate or nonexistent water treatment. The biggest user of water, agriculture, makes a substantial contribution. Nutrients and pesticides from crop and livestock management are major sources of *pollution*.

With no ongoing source of irrigation, rain-fed agriculture is dependent on rainfall to produce its crops. About 60% of the world's crop production is produced by rain-fed agriculture in a range of production systems (Table 3.1). The most productive systems are found in the humid and subtropical regions, as well as in temperate zones of Europe, North America, and the subcontinent. In dry tropic regions and highland areas, rain-fed cropping is typically associated with subsistence farming systems due to its relatively low yielding nature. Less than 30% of rainfall is utilized by plants during cultivation, according to data from farms around the world. The remainder enters rivers as runoff, seeps into groundwater, or evaporates into the atmosphere.

Question

What is dry farming?

Answer

Cultivation of crops in areas where rainfall is less than 750 mm per annum. Dryland Farming: Cultivation of crops in areas receiving rainfall above 750 mm. Rainfed Farming: Cultivation of crops in regions receiving more than 1,150 mm.

Keyword

Pollution is the introduction of harmful materials into the environment.

3.2.1. Types of Agricultural Regions

- **Subsistence Agriculture:** The production of food for the farmer and his family's consumption is referred to as subsistence agriculture. This kind of agriculture is prevalent in many developing nations, where small-scale farmers grow food for their own consumption using conventional farming methods.
- **Commercial Agriculture**: Producing livestock and crops for the market is known as commercial agriculture. This kind of agriculture is prevalent in developed nations, where large-scale farms generate food and raw materials for consumers and industries using contemporary technologies and farming methods.
- **Mixed Agriculture**: Combining commercial and subsistence farming techniques is known as "mixed agriculture." Small-scale farmers in developing nations often engage in this kind of agriculture, producing food for both market sale and personal consumption by utilizing both conventional and modern farming techniques.

3.2.2. Examples of Agricultural Regions

- **Southeast Asia:** Subsistence agriculture is prevalent in Southeast Asia, where small-scale farmers grow food for their own consumption by using conventional farming methods.
- **North America:** A sizable portion of commercial agriculture is found in North America, where large-scale farms employ cutting-edge techniques and technologies to produce food and raw materials for consumers and industry.

3.2.3. Issues in Agricultural Regions

- **Food Security**: Because access to food is restricted in many parts of the world due to poverty, conflict, and natural disasters, ensuring food security is a crucial issue.
- **Environmental Impacts**: Deforestation, soil erosion, and pollution are just a few of the negative environmental effects that can result from producing food and raw materials.
- **Economic Impacts**: Significant economic effects of agriculture can be both positive and negative, resulting in social inequality, rural-urban migration, and the creation of jobs and revenue.

3.2.4. Strategies for Addressing Agricultural Regions

- **Promoting Sustainable Agriculture**: Addressing the negative effects of agriculture on the environment and society requires promoting sustainable agriculture. This involves encouraging environmentally friendly farming methods including conservation agriculture, agroforestry, and organic farming.
- **Encouraging International Cooperation:** Promoting global collaboration is imperative in tackling the issues of food security and the financial difficulties confronting farming areas. This entails encouraging cooperation on trade, R&D, and agricultural policy among nations and regions.
- **Investing in Technology and Innovation**: Agricultural regions face many challenges that can only be solved by investing in innovation and technology. This entails creating fresh approaches to the production of crops, livestock, and food processing and distribution.

Did you get it?

1. **What is the irrigation by crop in the continental United States from 2008 to 2020?**
2. **What do you mean by agricultural region?**

3.3. WORLD AGRICULTURE TYPOLOGIES

Learning Objectives

- Discuss about fallow
- Perform crop rotation
- Explain mixed cropping
- Discuss about two-cropping agriculture
- Describe multi-cropping agriculture
- Know relay cropping
- Enhance crop-productivity
- Understand agricultural efficiency and crop intensity

The classification of agricultural systems based on different factors like soil, climate, technology, and socioeconomic conditions is referred to as agricultural typology. A holistic approach is needed to grasp agricultural typologies, considering the complex interplay of soil, climate, technology, and socioeconomic aspects. It also emphasizes the importance of sustainable and ethical farming practices that respect the unique characteristics of different communities and regions.

3.3.1. Typology of Agricultural Practices and Techniques

Cultivating land and raising livestock are practices referred to as agriculture. The cultivation of domesticated animals produced food surpluses that allowed people to live in cities, and this was a major factor in the rise of sedentary human civilization. Below is a discussion of the nine main agricultural methods and practices.

- *Fallow:* Constant cultivation causes the soil to lose its fertility. In order to restore the land to its original fertility, it is periodically left fallow for a certain amount of time.
- *Crop Rotation:* This procedure is used to either prevent fallow land from occurring or shorten the time that it does. Leguminous plants, which naturally fix nitrogen from the atmosphere, are the most widely grown crops in crop rotation.
- *Mixed Cropping:* In order for the nutrients used by one crop to be partially

replaced by the nutrients used by the other crops, this practice involves growing two or three crops concurrently on an agricultural plot. The ratio of crops to be mixed varies based on the needs of the area and the methods used.

- *Two-Cropping Agriculture:* This technique involves rotating the cultivation of two crops each year. In places with sufficient rainfall or suitable irrigation facilities, it is a common practice. Typically, the crops that fix nitrogen are grown as the second crop, e.g., crops made of legumes.
- *Multi-cropping Agriculture:* In areas where early maturing crop varieties and improved water management techniques are available, this practice—which typically involves growing three crops annually—is prevalent.
- *Relay Cropping:* Relay cropping is the process of sowing a new crop while the one that is ripening remains in the field.
- *Crop Productivity:* It speaks of the quantity of crop produced per worker or per hectare. While productivity per worker is high in areas of extensive agriculture, productivity per hectare is high in areas of *intensive agriculture.*

a. *Agricultural Efficiency:* It is focused on determining the highest quantity that can be produced in the shortest amount of time and sold for.

b. *Crop Intensity:* In nations like India, the amount of arable land is practically infinite. As a result, harvesting multiple crops from the land in a single year is imperative. Crop intensity refers to the best possible agricultural use of a piece of land.

Keyword

Intensive agriculture, also known as intensive farming (as opposed to extensive farming), conventional, or industrial agriculture, is a type of agriculture, both of crop plants and of animals, with higher levels of input and output per unit of agricultural land area.

Question

What is subsistence agriculture?

Answer

Subsistence agriculture is the production of food primarily for consumption by the farmer and is mostly found in less developed countries. In subsistence agriculture, small-scale farming is primarily grown for consumption by the farmer and their family. Sometimes if there is a surplus of food, it might be sold, but that is not common. In commercial agriculture, the primary objective is to make a profit.

Did you get it?

1. **Who presented the typology of world agriculture?**
2. **How many types of agriculture are there in the world?**

3.4. FACTORS AFFECTING AGRICULTURE

Learning Objectives

- Understand economic factors affecting farming
- Identify climatic factors affecting farming
- Explain physical/environmental factors affecting farming

For hundreds of millions of hungry and impoverished people worldwide, agriculture is their primary source of income. Approximately the size of an American football field or soccer pitch, the majority of them are connected to small farms or small farms. A small number of crops were grown in the early days of agriculture, and most food was obtained from the wild. Weather and soil changes might have helped agriculture continue to expand. Unlike hunting, agriculture can provide food for a larger population in the same area. Agriculture started out mostly for personal use but has since developed into a commercial industry.

Crop rotation, fencing, the use of fertilizer, plantations, weeding, livestock, and breeding are just a few of the many techniques that agriculture has come to encompass. This technology is intended to raise yields in agriculture by increasing production. Similarly, a number of factors have led to the widespread practice of agriculture in the modern world. Global agriculture is impacted by the following factors.

3.4.1. Economic Factors Affecting Farming

Agriculture is an ancient practice, but its modern form is influenced by a variety of economic factors.

- Subsidies: The government provides farmers with subsidies to aid in the growth of their crops. In the meantime, the government offers financial incentives to farmers for specific crop varieties.
- Commodity prices: The prices of major commercial crops like corn and soybeans are influenced by weather, investor speculation, and demand for edible and non-food crops (like biofuels). Depending on the general production prices that

industrial buyers are willing to pay, farmers can profit or lose money when using raw materials.

- Immigration laws and labor: Mainstream agriculture is centered around low-wage workers, namely immigrant farmers. Agriculture is impacted by immigration law because it has an impact on labor availability, which includes labor laws that determine whether or not to allow wages from subsistence farming.

3.4.2. Climatic Factors Affecting Farming

Agriculture is influenced by various climate-related factors such as light, water, rain, temperature, air quality, relative humidity, and wind.

- Light: Plants need light to produce chlorophyll and carry out photosynthesis. Light also affects processes like photomorphogenesis, abscission, stomatal movement, *translocation*, mineral absorption, and phototropism.

Keyword

Translocation is a type of chromosomal abnormality in which a chromosome breaks and a portion of it reattaches to a different chromosome.

- Water and rainfall: Water is essential for both plant and animal life. Plant growth, development, and yield are all influenced by the availability of water.
- Temperature: It affects many crop growth processes, such as photosynthesis, transpiration, respiration, protein synthesis, translocation, germination, and seed dormancy disruption.
- Air: Plants need to breathe in order to produce the energy required for various growth and development processes. Carbon dioxide is essential for plants to produce food during photosynthesis.
- Relative humidity: The maximum amount of water vapor that the air can hold depends on its temperature. Cold air has a lower capacity for water vapor compared to warm air.
- Wind: Air is moved by wind because of different pressure gradients and warming. Fruit and seed development is aided by pollination, which is facilitated by air. Gas exchange occurs in plants in mild winds. On the other hand, high winds have the potential to encourage crop toppling and water loss.

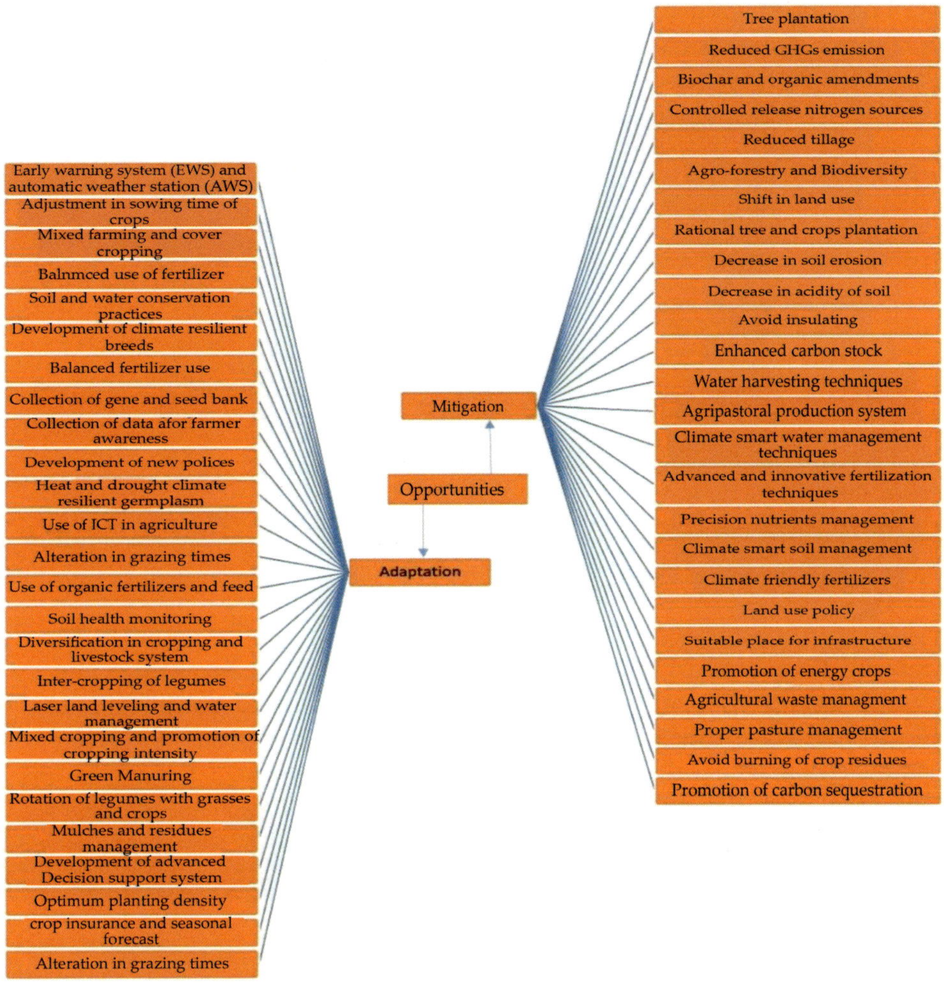

Figure 3.2. Factors affecting farming.

Source: Habib-ur-Rahman, M., Ahmad, A., Raza, A., Hasnain, M. U., Alharby, H. F., Alzahrani, Y. M., Bamagoos, A. A., Hakeem, K. R., Ahmad, S., Nasim, W., Ali, S., Mansour, F., & El Sabagh, A. (2022). Impact of climate change on agricultural production; Issues, challenges, and opportunities in Asia. *Front. Plant Sci. 13*, 925548. doi: 10.3389/fpls.2022.925548.

3.4.3. Physical/Environmental Factors Affecting Farming

A number of environmental factors have an impact on agriculture. The three primary physical factors that impact agriculture are topography, soil, and climate (Figure 3.3).

- Topography/relief: Poor transportation infrastructure, soil erosion, and challenges with land cultivation are all related to topography. Depending on the land's topography, agriculture may be mechanized.
- Climate: Climate factors such as precipitation and temperature affect agriculture as described above.

- Soil: The plant does well in well-drained, nutrient-rich loamy soil. Plants utilize their roots to absorb nutrients and water from the soil. Soil that is poorly textured and has been exposed to harsh chemicals yields less.

Figure 3.3. Locust swarm near Satrokala, Madagascar in 2014.

Source: By Iwoelbern – Own work, CC BY-SA 3.0, https://commons.wikimedia.org/w/index.php?curid=33198217.

Farmers must be aware of these factors in the specific area where they plan to grow so that they can select the best crop to grow. The secret to excellent crop growth and yield is selecting the appropriate crop for each farming region.

Question

What are the factors affecting agricultural supply?

Answer

Those factors include the price of the product in question, the number of producers, the input costs, the technological changes, the price of other possible products, and unpredictable factors such as weather. The relationship between quantity supplied and price can be described by the elasticity of supply.

Did you get it?

1. **What is the most important factor in agriculture?**
2. **What are the factors affecting agricultural mechanization?**

3.5. FOOD SECURITY

Learning Objectives

- Explain the nature of food insecurity: duration, severity and types
- Describe trends and patterns in food insecurity
- Identify causes and geography of famine
- Focus on key dimensions of food security
- Give efforts and remedies to ensure food security

According to the World Food Summit of 1996, food security is achieved when all people have physical and financial access to enough safe, nutrient-dense food that satisfies their dietary needs and food preferences for an active and healthy life, regardless of their circumstances.

3.5.1. Main Dimensions of Food Security

The four main dimensions of food security are as follows:

- Physical *availability* of food: Food production, stock levels, and net trade all affect food availability, which deals with the "supply side" of food security.
- Economic and physical *access* to food: Food security at the household level is not assured by a sufficient supply of food on a national or international scale. A growing policy focus on incomes, expenditures, markets, and prices in order to achieve food security goals has resulted from concerns about inadequate food access.
- Food *utilization*: The process through which the body maximizes the different nutrients present in food is commonly referred to as utilization. A person's ability to consume enough energy and nutrients depends on a variety of factors, including proper care and feeding techniques, food preparation, a varied diet, and the distribution of food within the household. When combined with optimal biological utilization of ingested food, this establishes an individual's nutritional status.
- *Stability* of the other three dimensions over time: Even if your intake of food is sufficient right now, you could still be classified as food insecure if you occasionally have insufficient access to food, which could lead to a decline in your *nutritional status*. Your food security status may be impacted by unfavorable weather, unstable political environments, or economic factors like rising food prices and unemployment.

For food security objectives to be realized, all four dimensions must be fulfilled simultaneously.

3.5.2. Effects of Food Insecurity

Food insecurity is the root cause of both hunger and famine. A high level of susceptibility to famine and hunger is a result of chronic food insecurity, and achieving food security requires removing that susceptibility. People who are food insecure may be forced to engage in dangerous economic ventures such as prostitution.

Those who live in "food deserts," or areas where nutrient-dense foods are either inaccessible or prohibitively expensive, are also more likely to be food insecure and obese. Due to their frequent need to eat less nutrient-dense but more easily accessible food, residents of these neighborhoods are more likely to suffer from obesity, diabetes, and heart disease.

3.5.2.1. Stunting and Chronic Nutritional Deficiencies

Food distribution issues and shortages persist in many nations. These cause a considerable number of people to experience chronic and frequently widespread hunger. In response to prolonged starvation and malnourishment, human populations may experience stunting or stunted growth, which is characterized by a reduction in body size. If the mother is undernourished, this process begins in utero and lasts until the child is about three years old. Although the rates of infant and child mortality are higher than during famines, they are still much lower. After the age of roughly two years, increased nutritional intake cannot undo the effects of stunting. Early-life severe malnutrition frequently results in abnormalities in cognitive development. As a result, there is a difference between children who suffer from severe malnutrition and those who do not.

In 2019, there were 144 million children worldwide affected by stunting, which accounts for 21.3% of all children. Central Asia, Eastern Asia, and the Caribbean are the only subregions on track to meet the 2025 and 2030 stunting targets, showing the highest reduction rates in stunting prevalence. Globally, the prevalence of stunting in children decreased by one-third between 2000 and 2019. According to the 2021 FAO State of Food Security and Nutrition in the World (SOFI), in 2020, 22.0% of children under five suffered from stunting, 6.7% had wasting (45.4% of the population), and 5.7% were overweight. FAO warned that due to the impact of the COVID-19 pandemic, these numbers might be even higher. Over 90% of children worldwide experience stunting, over 90% have wasting, and over 70% are affected by being overweight. These statistics are mainly seen in Africa and Asia.

3.5.2.2. Mental Health Outcomes

Among the socioeconomic factors influencing mental health is food insecurity. Food insecurity is strongly linked to an increased risk of depression, anxiety, and sleep disorders, according to a recent thorough systematic review of over 50 studies. Those who are food insecure are nearly three times more likely to experience depression and anxiety than those who are food secure. Food insecurity has also been connected in studies to a higher risk of disordered eating patterns.

3.5.3. Strategy

By boosting "nutrition-sensitive agriculture," enhancing food safety, and enhancing food security, the World Bank Group collaborates with partners to create food systems that can feed everyone, everywhere, every day. The Bank is a top provider of financing for food systems. IBRD/IDA committed US$9.6 billion to agriculture and related sectors in the fiscal year 2022.

> **Keyword**
>
> **Nutritional status** has been defined as an individual's health condition as it is influenced by the intake and utilization of nutrients.

Activities include:

- Strengthening safety nets to ensure that vulnerable families have access to food and water–and money in their pockets to make vital purchases
- Delivering expedited emergency support by fast-tracking financing through existing projects to respond to crisis situations
- Addressing issues related to food security through collaboration with nations and development partners. The Famine Action Mechanism and the Agriculture Observatory are two examples of partnerships and data-based monitoring tools, along with quick country diagnostics.
- Encouraging farming practices that employ climate-smart methods and yield a wider variety of foods will strengthen the resilience of food systems, boost farm incomes, and make nutrient-dense foods more widely available and reasonably priced.
- Improving supply chains to reduce post-harvest food losses, improve hygiene in food distribution channels, and better link production and consumption centers.
- Applying an integrated "One Health" approach to managing risks associated with animal, human, and environmental health.
- Supporting investments in research and development that enable increasing the micronutrient content of foods and raw materials.
- Advocating for policy and regulatory reforms to improve the efficiency and integration of domestic food markets and reduce barriers to food trade.

- Working with the private sector, government, scientists, and others to strengthen capacities to assess and manage food safety risks in low and middle-income countries.
- The Bank houses the Global Agriculture and Food Security Program (GAFSP), a global financing initiative that brings together donor funds and provides additional funding to support agricultural development across the entire value chain. GAFSP was launched in 2010 by the G20 in response to the food price crisis of 2007–2008. Since then, it has provided over $1.3 billion in grant funding to 64 projects in 39 countries, benefiting more than 13 million smallholder farmers and their families. Additionally, GAFSP has allocated $330 million to 66 agribusiness investment projects in 27 countries and $13.2 million in small-scale grants to support producer organizations. In light of the ongoing COVID-19 pandemic, GAFSP has recently provided over $55 million in additional grant funding to support COVID-19 response and recovery in existing public sector and producer organization-led projects.
- Additionally, the Bank funds the CGIAR, which promotes agricultural innovation and science to increase food and nutrition security worldwide.

Question

What is wage?

Answer

Payment made to labor for its assistance to production is called wage.

Did you get it?

1. **What happens to marginal utility when total utility increases?**
2. **What are the dimensions of food security?**

CASE STUDY

YOUTH ENTREPRENEURSHIP FOR RESILIENCE IN THE CENTRAL AFRICAN REPUBLIC

The Central African Republic has experienced a series of crises that have lowered living standards, destroyed social and educational infrastructure, and undermined public and private institutions. Young people are especially affected because they frequently lack basic life skills and are easily recruited by armed militias if they are not assimilated into society (UN OCHA 2023b). In light of this, the initiative Agricultural Vocational Training for Unemployed and Internally Displaced Youth seeks to lessen migration and crime while assisting young people in the Bangui region between the ages of 18 and 35 in leading more socially and economically stable lives. In addition to providing training in various agricultural fields, the project also assists participants in launching their own enterprises, enabling them to successfully integrate into local food and economic systems.

After enrolling in the agricultural vocational training center for poultry farming in 2019, Nazaire Namkomana has completed his studies. His newfound abilities will allow him to support his family financially and nutritionally. “My family's needs were my top priority before the training; there were times when we went without food,” he says. “My kids' future education is now guaranteed. Despite my passion for the industry, I knew nothing about poultry farming and had no intention of doing it. After the training, I began managing a flock of fifty chicks, then grew it to one hundred, then to 150, and now I've placed an order for 200.

Following his training, Nazaire imparted his knowledge to other young people, saying, “My goal is to grow my unit into a sizable production company so that I can not only take care of my family and send my kids to prestigious schools, but also lessen poverty among the youth in my community.”

FOCUS ON CAREERS

AGRONOMIST

Agronomists are crop and soil scientists who strive to increase agricultural soil productivity in order to guarantee greater crop yields. They research all aspects of plants and agriculture, including cultivation, growing, genetic modification, handling, and applications for the betterment of society. Enhancing the quality and yield of crops is the main goal of agronomists. These experts frequently experiment on various plants in an effort to increase crop yield, longevity, and durability.

What does an agronomist do?

Agronomists can focus on research, sustainable development, soil and water conservation, crop management, and other areas. The particular responsibilities of an agronomist differ according to their employer, but typical duties consist of:

- Conducting crop experiments in a lab
- Meeting with farmers to discuss possible farm improvements
- Collaborating with other researchers to produce ideas for plant development
- Solving problems related to planting, harvesting, storage, and distribution
- Determining methods for protecting plants from harsh weather or climate
- Traveling around the country to visit farms and meet with their owners
- Presenting research at conferences or publishing it in journals
- Developing ideas for sustainable and environmentally conscious farming practices
- Engineering methods for conserving and recycling water and other resources

Agronomist Career Requirements

There are a few prerequisites you should take into account before pursuing a career in agronomy. Initially, a formal education in agriculture is a prerequisite for becoming an agronomist. Most agronomists decide to enroll in an accredited college or university to pursue a bachelor's degree in agriculture. Specific courses on agronomy and agronomic research are offered in many programs. Generally speaking, a master's degree is necessary for agronomists who want to focus on research. Agronomists need to be able to physically inspect farms in a range of climates and environments. This involves the capacity to stoop, kneel, and lift modest objects. Agronomists are also adept at critical thinking, teamwork, and communication.

How to become an agronomist

Here are the typical steps an aspiring agronomist should follow to pursue this career:

1. Earn a bachelor's degree

A bachelor's degree in agriculture is the first requirement for becoming an agronomist. Many of the nation's colleges and universities provide in-person and virtual courses that teach students how to identify and study plants, work through issues pertaining to agricultural development, and carry out scientific trials and experiments.

2. Get a master's degree

Subsequently, the majority of agronomists typically submit applications and join a master's degree program. Agronomists keen on research or academia find Master of Science degrees especially advantageous. The coursework at the graduate level delves into various topics in agronomy such as plant genetics, biochemistry, physiology, and crop management.

3. Find an internship

Agronomists have the opportunity to apply for internships that provide the chance to gain real-world experience and mentoring post-graduation. Internship opportunities are available in agribusiness firms, educational institutions, and federal agencies. The majority of internships involve supporting experienced agronomists on farms or as laboratory research assistants.

4. Apply for jobs

After completing their degree and internship, agronomists can begin seeking employment. They often find job opportunities through industry databases, educational networks, and online job boards. Additionally, agronomists need recommendation letters from professors or previous employers, a professional cover letter, and a resume.

5. Build your credentials

Agronomists can enhance their credentials by obtaining three main certifications: Certified Crop Advisor, Certified Professional Agronomist, and Certified Professional Soil Scientist/Classifier. These certifications not only improve job prospects but also demonstrate commitment to the field. To become certified, agronomists must meet educational requirements and have two years of industry experience. Most certification bodies also require agronomists to participate in continuing education programs. Joining professional organizations like the Crop Science Society of America is another effective way to strengthen one's credentials as an agronomist.

Average Salary of An Agronomist

The annual salary of an agronomist is $69,303 on average. Salary levels are influenced by a person's location, level of education, and credentials. Based on how well they perform at work, agronomists may be promoted or given raises. For example, the average annual salary of an agronomy manager is $82,426.

Agronomist Career Outlook

From 2021 to 2031, the total employment of agricultural and food scientists is expected to grow by 8%, according to the US Bureau of Labor Statistics. Agronomists should anticipate reasonable job security because this growth is outpacing the average for all occupations. Career opportunities in the field of research and development are anticipated to be abundant and diverse for agronomists upon graduation. Academic and educational opportunities might be more scarce.

According to the *Bureau of Labor Statistics* (BLS), a career in agronomy is more stable than it has been recently for many other scientific occupations. In the upcoming ten years, there shouldn't be many job losses in the sustainable industries of crop development and food production. Agronomists usually find it easy to expand their professional networks and look for employment opportunities because they operate in a variety of environments.

Where do agronomists work?

Agronomists can work in a variety of settings, depending on their field of study. The majority of an agronomist's time is spent either on a farm or in a lab. Agronomists have laboratories stocked with live plants for experiments, soil samples, and scientific instruments. When working in a lab, agronomists typically wear gloves, protective coats, and eyewear. Agronomists usually dress casually when visiting farms, covering up with sunblock such as hats and sunglasses. In greenhouses, orchards, and fields, agronomists inspect plants. They might make frequent trips to farms in other states or nations to meet with farmers. An agronomist may present their research at a conference for science or academia after conducting both theoretical and practical research.

A CLOSER LOOK

CLIMATE SMART AGRICULTURE

Climate-smart agriculture is an approach to livestock production and crop cultivation that takes the present and potential effects of climate change into account. It entails applying sustainable farming techniques that assist farmers in maintaining or increasing their production levels while assisting in the reduction of greenhouse gas emissions. Farmers can also benefit from climate-smart agriculture by increasing crop yield and evaluating future crop yield and natural disaster risks. The primary objective of climate-smart agriculture is to support farmers in meeting global food needs while safeguarding the environment and mitigating the effects of climate change.

It is difficult for farmers in the majority of developing nations to implement climate-smart agriculture, though. Using climate-smart agriculture may present some difficulties for farmers, such as:

1. Low knowledge and awareness: Many farmers might need to learn about and become familiar with climate-smart agriculture techniques to apply them to their farms. Even if farmers are interested in climate-smart agriculture, it is challenging for them to make the switch, particularly in developing nations.
2. High upfront costs: Many farmers may find it prohibitively expensive to implement some climate-smart agriculture techniques since they call for large investments in new infrastructure or technology. Furthermore, adopting precision farming methods might necessitate a farmer investing in staff training or new equipment purchases. Nonetheless, long-term crop yield optimization can be achieved by farmers with support for climate-smart agriculture.
3. Market challenges: Sometimes, especially when there is little market demand, farmers may require assistance in locating buyers for the climate-smart agricultural products they produce. Making the switch to climate-smart agriculture practices financially feasible and recovering their investments is a challenge for farmers.
4. Policy and regulatory barriers: In general, policy and regulatory barriers that impede the adoption of climate-smart agriculture practices may pose difficulties for farmers. For instance, there might be limitations on the application of particular procedures or technologies, or government organizations might not provide enough financial or technical support.
5. Social and cultural barriers: Agriculturalists may encounter social and cultural obstacles that make the adoption of climate-smart practices difficult. If farmers are unfamiliar with or at odds with traditional farming practices, they might be reluctant to embrace new technologies or procedures. In addition, farmers who try to apply disruptive or contentious climate-smart agriculture practices may encounter opposition from the local community.

The degradation of natural resources is one of the primary causes of the rapidly rising cost of food. Implementing climate-smart agriculture, increasing yields, and ensuring food security are some ways to combat this. Plantations in Southeast Asia, which make major contributions to the production and export of commodities worldwide, are an ideal place to begin. Additionally, there are a number of benefits for Southeast Asian plantations that implement climate-smart agriculture.

- Increased productivity and yield: Climate-smart agriculture techniques maximize the use of water, nutrients, and other inputs, which can help to increase crop yields and productivity.
- Improved water management: Climate-smart agricultural techniques can minimize water loss and increase irrigation effectiveness, enabling farmers to more effectively manage their water resources.
- Enhanced soil health: The health of the soil can be enhanced by using climate-smart agricultural techniques like cover crops and conservation tillage.
- Reduced greenhouse gas emissions: The implementation of climate-smart agriculture practices has the potential to mitigate greenhouse gas emissions, enhance the efficacy of water and nutrient management, and decrease the reliance on fossil fuel-derived inputs like fertilizers and pesticides.
- Improved farmer resilience: By arming farmers with the information and resources they need to adjust to shifting weather patterns and other difficulties, climate-smart agriculture practices can help them become more resilient to the effects of climate change. It promotes the long-term viability of plantation agriculture in Southeast Asia.

SUMMARY

- Cultivating natural resources for human sustenance and financial benefit is what we call agriculture. It blends new technologies and contemporary production techniques with the ingenuity, imagination, and skill required to raise animals and plant crops.
- The output of agriculture, which includes raw material production, participation in the global supply chain, and economic development, is a major factor in the industry's and society's importance.
- The technology sector is given significant attention when disruptions occur in global supply chains. For instance, a manufacturer's ability to produce computers, automobiles, and other goods is hampered by the lack of computer chips, which are made from silicon, a non-agricultural raw material. This affects numerous facets of business and society.
- Agriculture has played a crucial role in daily life for thousands of years. Before the advent of agriculture, humans relied on hunting and gathering for survival. Humanity didn't start to flourish until it switched to the scheduled planting and harvesting of crops.
- Crop yields have increased as a result of improved farming machinery and increased use of pesticides and fertilizers.
- The ability to produce enough food to feed a growing population and the amount of land that can be used for agriculture have always required careful balance.
- Addressing the negative effects of agriculture on the environment and society requires promoting sustainable agriculture. This includes promoting environmentally friendly farming methods such as conservation agriculture, agroforestry, and organic farming.
- The categorization of agricultural systems according to various parameters, including soil, climate, technology, and socioeconomic circumstances, is known as agricultural typology.
- Crop rotation, fencing, the use of fertilizer, plantations, weeding, livestock, and breeding are just some of the various techniques that agriculture now includes. This technology is intended to raise yields in agriculture by increasing production.
- Agriculture is influenced by various climate-related factors such as light, water, rain, temperature, air quality, humidity, and wind. Topography, soil, and climate are the main physical factors affecting agriculture.
- Food security is achieved when all people have physical and financial access to enough safe, nutrient-dense food that satisfies their dietary needs and food preferences for an active and healthy life, regardless of their circumstances.
- Food insecurity is the main cause of hunger and famine. A high vulnerability to famine and hunger stems from chronic food insecurity, and achieving food security involves eliminating that vulnerability. Individuals facing food insecurity may be compelled to resort to risky economic activities such as prostitution.

- By boosting "nutrition-sensitive agriculture," enhancing food safety, and improving food security, the World Bank Group collaborates with partners to create food systems that can feed everyone, everywhere, every day.

MULTIPLE CHOICE QUESTIONS

1. **Mines and Fishes are subject to the law of:**
   a. Diminishing return
   b. Increasing return
   c. Constant return
   d. None of the above
2. **Which of these statements is correct?**
   a. Agriculture economics may be taken as the branch of both agriculture and economics
   b. Agriculture economics is more intimately related to economics than to agriculture
   c. Agricultural economics is nothing but the application of economic principles to agriculture
   d. All of these
3. **When we study a problem from particular to general, the method is known as?**
   a. Deductive method
   b. Inductive method
   c. Both (a) and (b)
   d. None of the above
4. **Name the market in which permanent or durables commodities are traded:**
   a. Long period market
   b. Short period market
   c. Secular market
   d. Bullion market
5. **Net capital ratio is equal to:**
   a. Total assets/Total liabilities
   b. Total liabilities/Total assets
   c. Total assets/Current liabilities
   d. None of the above
6. **Net return per hectare is:**
   a. Overall efficiency measure
   b. Partial efficiency measure
   c. Specific efficiency measure
   d. None of the above

7. When variable cost is zero, the total cost will be?

a. Equal to variable cost

b. Equal to fixed cost

c. Equal to average variable cost

d. None of the above

8. Opportunity cost is:

a. Cost of supplementary enterprise

b. Cost of next best alternative foregone

c. Cost of cultivation

d. Cost of production

9. Optimizing the use of farm resources on an individual farm level. It is a:

a. Farm management

b. Production economics

c. Agricultural marketing

d. Macro economics

REVIEW QUESTIONS

1. What are agriculture typologies?
2. Discuss the various agricultural typologies prevalent around the world, focusing on factors such as climate, soil, technology, and socio-economic conditions.
3. Examine the concept of agricultural typology and how it has evolved over time. Analyze the implications of these changes on sustainability and food security in both the Global North and the Global South.
4. How do oxygen and carbon dioxide affect agricultural production? Explain.
5. Explain the significance of soil quality, including soil texture and nutrient content, in agriculture and how it can be managed.
6. How does access to water resources, such as irrigation, impact agriculture in arid and semi-arid regions?

Answer to Multiple Choice Questions

1. (a)	2. (d)	3. (b)	4. (c)	5. (a)
6. (a)	7. (b)	8. (b)	9. (a)	

REFERENCES

1. Ball, V. E., Butault, J. C., Mesonada, C., & Mora, R., (2010). Productivity and international competitiveness of agriculture in the European Union and the United States. *Agricultural Economics, 41*(6), 611–627.
2. Butzer, R., Mundlak, Y., & Larson, D., (2012). Measures of fixed capital in agriculture. In: Fuglie, K., Wang, S. L., & Ball, V. E., (eds.), *Productivity Growth in Agriculture: An International Perspective*. Wallingford, UK: CAB International.
3. Cahill, S., & Rich, T., (2012). A production account for Canadian agriculture, 1961–2006. In: Fuglie, K., Wang, S. L., & Ball, V. E., (eds.), *Productivity Growth in Agriculture: An International Perspective*. Wallingford, UK: CAB International.
4. Cramer, G. L., (2010). *Agricultural Economics*. Milton Park, Abingdon, OX: Routledge.
5. Desai, P. K., (2010). *Agricultural Economics*. Delhi: Biotech Books.
6. Drummond, H. E., (2011). *Agricultural Economics* (3rd edn.) Boston: Prentice Hall.
7. Evenson, R., & Fuglie, K., (2010). Technology capital: The price of admission to the growth club. *Journal of Productivity Analysis, 33*(3), 173–190.
8. Kováts, I., (2010). *Economics of Sustainable Agriculture*. Gödöllö: SzentIstván University.

CHAPTER 4

Industry

LEARNING OBJECTIVES

After studying this chapter, you will be able to:

- Understand an industry
- Identify factors affecting the location of industries
- Focus on industrial development and economic growth

"In the ordinary business of life, industry can do anything which genius can do, and very many things which it cannot."

—Henry Ward Beecher

INTRODUCTORY EXAMPLE

Industrial Economics

A study of firms, industries, and markets is known as industrial economics. Businesses of all kinds are examined, from neighborhood corner stores to global behemoths like Tesco and Walmart. Additionally, it takes into account a wide variety of industries, including the manufacturing of cars, restaurants, and power.

When analyzing decision-making at the levels of the individual firm and industry, Industrial Economics helps us understand such issues as:

- the levels at which capacity, output, and prices are set;
- the extent to which products are differentiated from each other;
- how much firms invest in research and development (R&D);
- how and why firms advertise.

In addition to examining motivation, industrial economics provides insights into how businesses structure their operations. While maximizing profits is often assumed in micro-courses, many industrial economics courses look at other goals, like trying to increase market share. There's also the global aspect: companies can choose to outsource production or source inputs from other countries. As such, macroeconomic concepts are occasionally used in industrial economics, but the field more often uses skills and knowledge from micro-courses.

Determining whether a market is competitive is one of the main concerns in industrial economics. The majority of industrial economics courses analyze how to gauge the level of market competition because competitive markets are typically advantageous for consumers (though they may not always be practical). Subsequently, it evaluates the necessity of regulation and the appropriate form for it. Once more, there is an international component to this because businesses that operate internationally will encounter various regulatory frameworks.

As a result, students should be prepared to use diagrams and possibly some simple mathematical models, such as game theory, to understand firm and regulatory decision-making in industrial economics. Furthermore, academics frequently create empirical statistical models to find correlations between relevant variables, such as the relationship between profits, advertising, and product price. Even though econometrics courses handle conducting independent empirical analysis, most other courses do not mandate it, comprehension and interpretation of empirical results are crucial abilities.

Moreover, industrial economists are highly employable. The Office of Fair Trading (OFT) and the Competition Commission (CC) are two government organizations and consultancies that are involved in competition policy. An equally large number of consultancies and regulators, including Ofcom, the regulator for the communication sector, are concerned with the economics of regulation.

INTRODUCTION

An industry is a collection of businesses connected by their main lines of business. There are numerous industry classifications in modern economies. Usually, industry classifications are arranged into more comprehensive groups known as sectors. An industry is typically used to group together individual businesses according to their main sources of income. For instance, most classification systems would place an automobile manufacturer in the automaker industry even though their financing division accounts for 10% of their total sales.

4.1. UNDERSTANDING AN INDUSTRY

Learning Objectives

- Understand the classification of industry
- Discuss the distribution of major industries
- Focus on Global Industry Classification Standard (GICS)
- Describe the advantages of basic industry
- Identify limitations of basic industry

The main product that similar businesses manufacture or sell is what groups them into industries. This gives rise to industry associations, which subsequently serve as a means of distinguishing companies from various other groups. Economists and investors frequently research industries to gain a better understanding of the drivers and constraints surrounding the expansion of corporate profit. To determine how attractive a company is in comparison to others in the same industry, companies can also be compared to one another.

Because the same macroeconomic factors affect every member of an industry, stocks within it frequently rise and fall together. These macroeconomic variables may include shifts in investor sentiment in the market, such as a reaction to a specific news story or event. There may also be modifications aimed at the industry as a whole, like new rules or higher raw material prices. However, developments of a single company may result in the related stock rising or falling independently of other stocks in the same sector. Events like the introduction of a unique product, a public corporate scandal, or a shift in the organization's leadership structure may cause this.

Did you know?

Stocks of companies operating within the same industry tend to have similar stock price movements. This is because companies within a given industry face the same headwinds, challenges, and broad triumphs.

4.1.1. Classification of Industry

4.1.1.1. Size

The size of an industry is measured by the amount of money invested, goods produced, and number of employees involved. Industries can be classified as either large or small. Modern technology and substantial capital expenditures are common in large-scale industries. Instances include the oil and gasoline sector or heavy machinery. The contrast between large- and small-scale industries is pronounced. Their technology is not advanced, and they need less capital. Additionally, manual labor is used more often in this kind of industry. Pottery and weaving are a few examples of this industry.

4.1.1.2. Raw Material

The raw materials can be separated according to industries. There are, for instance, industries related to agriculture, the sea, minerals, etc. Plants and animal products are the primary raw materials used in agro-based businesses. For instance, the dairy, leather, *textile*, and cotton industries, etc. Seawater or the ocean provides the raw materials used in marine-based industries. Fish oil, for instance. Conversely, businesses that rely on minerals are built on mining and typically use mineral ore as their primary raw material. Additionally, these raw materials can be used to make heavy machinery or building materials.

Keyword

A textile is a flexible material made by creating an interlocking network of yarns or threads, which are produced by spinning raw fibers into long and twisted lengths.

4.1.1.3. Ownership

An industry may be jointly owned, privately owned, or owned by the public. In the private sector, an individual or group of individuals manages and owns the businesses and operations. In joint sector industries, the company is jointly owned by the state and an individual, such as Maruti Udyog. In the public sector, the industry is owned and managed by the government. An example is the Oil and Natural Gas Corporation (ONGC).

4.1.2 Distribution of Major Industries

Industries were categorized according to raw material, size, and ownership. This section will provide information on the distribution of some important industries. The iron and steel and textile industries are among the oldest sectors contributing to Indian industrialization. Information technology is a newer field.

Countries such as Germany, the United States of America, China, Japan, and Russia have strong industries related to iron and steel. Meanwhile, the textile industries in South Korea, Hong Kong, and India are experiencing continuous growth. Silicon Valley in California and Bangalore in India serve as the hubs of the emerging information technology sector.

4.1.2.1. Iron and Steel Industry

The iron and steel sectors are renowned for being the backbone of all other industries, providing raw materials for various sectors. These industries involve diverse inputs, processes, and outputs, as we have studied in the industrial system. Inputs include labor, capital, infrastructure, and raw materials like iron ore. Smelting and refining are key processes in converting iron ore into steel.

Ultimately, steel is manufactured. Iron and steel are considered essential materials that support all other industries, forming the cornerstone of modern industry. India, a developing nation, benefits from the iron and steel industry due to low labor costs, abundant raw materials, and an open market.

4.1.2.2. Textile Industry

A textile is a fabric created through weaving fibers. Through a process known as spinning, raw materials like cotton or wool are transformed into yarn, which is then used to make fabric. Artificial or natural fibers are both possible. Cotton, jute, linen, wool, and silk are examples of natural fibers. Nylon, rayon, and polyester are examples of man-made fibers.

Since ancient times, people have worn and used textiles. One of the oldest industries in the world is the textile sector. Additionally, the textile industry wove fiber using wheels and looms prior to the Industrial Revolution. Power looms were initially introduced in Britain during the revolution.

Subsequently, Mumbai's textile industry flourished due to the city's warm, humid climate, its port for importing machinery and exporting finished goods, and most importantly, its plentiful inexpensive labor. Several highly sought-after and renowned fibers include gold-wrought cotton from Surat, Burhanpur, and Vadodara, chintzes from Masulipatnam, and calicos from Calicut.

4.1.2.3. Information Technology

The distribution, processing, and storing of information are all covered by information technology. A number of political, technological, and socioeconomic developments over the past ten years have brought the industry international attention. Information technology hubs are emerging in Bangalore, Hyderabad, Chennai, Mumbai, and other cities in India.

In terms of the advancement of information technology, Silicon Valley and Bangalore are similar in many ways, including having a temperate climate, a workforce with a high level of education, access to markets, and high-caliber scientific and technological centers.

4.1.3. Industries vs. Sectors

Systems of classification are used to group related business operations into sectors and industries. But sectors cover more ground than industries do.

For example, retail trade is a sector within the North American Industry Classification System (NAICS). Within that sector are industries, such as health and personal care stores, clothing stores, and shoe stores. Rite Aid Corporation and Gap, Inc. are members of the same consumer goods sector, but each would be listed in a different industry based on the specifics of the products they produce or sell. Rite Aid Corporation is classified within the health and personal care stores (NAICS Code 44610), while Gap, Inc. is classified within both the clothing stores industry (NAICS Code 448130) and clothing accessories industry (NAICS 448150).

Keyword

A **consumer** is a person or a group who intends to order, or use purchased goods, products, or services primarily for personal, social, family, household and similar needs, who is not directly related to entrepreneurial or business activities.

It is noteworthy that a single company may operate in two distinct sectors or industries. Rite Aid is regarded as a component of the personal services sector within the photofinishing laboratories industry in addition to the consumer goods sector. This company is given multiple NAICS codes (including NAICS Code 812921 for its photo department) because it produces photographs in addition to selling consumer goods.

When compiling statistical data, government agencies classify businesses using the North American Industry Classification System (NAICS), which was developed by the United States, Canada, and Mexico. Businesses in the same industry are classified under the same NAICS category if they employ comparable production methods.

4.1.4. Global Industry Classification Standard (GICS)

Another classification scheme that is frequently used as a reference is the Global Industry Classification Standard (GICS). Every public company is assigned by GICS to an industry group and economic sector that best describes its business. Standard and Poor's (S&P) and Morgan Stanley Capital International (MSCI) collaborated to develop the GICS in 1999. It was designed to be an effective investment tool that could track the evolution, depth, and breadth of different industry sectors. Investors, analysts, economists, and the MSCI indexes use the GICS methodology to evaluate and contrast rival businesses.

The GICS is a hierarchical, four-tiered system of industry classification. There are eleven economic sectors, based on the GICS hierarchy. Twenty-four industry groups, sixty-nine industries, and one hundred fifty-eight sub-industries are further subdivided into these sectors. Every stock at each of these four levels can be uniquely identified by its code. An example of an economic sector is "materials." Different industries fall under the category of materials, including those that deal with chemicals, building supplies, packaging, metals and mining, paper goods, and forest products.

Question

What is an example of an industry?

Answer

The NAICS classifies 'Finance and Insurance' as its own top-level sector. Within this sector, there are a variety of groups such as credit intermediaries, financial investment companies, insurance carriers, funds, trusts, and other financial vehicles. A specific NAICS industry is Commercial Banking, and it is assigned its own classification code (522110).

4.1.5. Advantages of Basic Industry

The basic industry has many advantages for a country and its citizens, including:

- Providing stable, well-paying jobs for a large portion of the population, as these industries often require significant amounts of labor.
- Generating government revenue through taxes, which can then be used to fund other programs.
- Offering a range of goods and services that are essential to the everyday lives of citizens, from food items to transportation.
- Stimulating growth in other industries, as the success of basic industries will often lead to the growth of related sectors.
- Helping to diversify a country's economy and reduce its reliance on a single sector or market.
- Creating a stable source of resources that can be used to manufacture goods or facilitate trade.
- Providing a platform for the development of new technologies, as well as the development of a skilled labor force.

4.1.6. Limitations of Basic Industry

An economy's core industry serves as the basis for all other industries and sectors, making it a crucial part of any given nation. It does, however, have certain drawbacks that could seriously jeopardize the nation's overall economic stability.

- High capital requirements: The high capital requirements associated with basic industries can make it difficult to establish and operate these businesses, as well as create an uncertain economic environment.
- Limited access to resources: Because of the nature of the industry, it is often difficult to access resources necessary for the production of goods and services. This can lead to cost increases and ultimately limit the industry's growth.
- Difficult to innovate: Innovation is important for any industry to remain competitive and successful; however, the basic industry often lacks the necessary resources

to innovate and develop new products and services.

- Increased competition: The basic industry often faces more competition from domestic and foreign competitors, which can lead to price wars and lower profits.
- Social and environmental impacts: The basic industry can have negative effects on the environment, such as water and air pollution, as well as social and economic impacts, such as displacement of communities.

Industry is a classification term used in business and investing to group companies that are generally similar. These businesses offer comparable goods and services, and they frequently react to macroeconomic developments in similar ways. Their stock prices frequently rise or fall simultaneously as a result. The Global Industry Classification Standard and the North American Industry Classification System are two widely used industry classification schemes.

Sectors are a more comprehensive classification that can contain several industries, and they can be further divided into industries. For instance, the home goods and apparel sectors can both be included in the retail sector. Certain macroeconomic trends may not affect businesses in the same sector in the same way.

Keyword

Profit is the benefit realized when the amount of revenue gained from an activity exceeds the expenses, costs, and taxes needed to sustain the activity.

Did you get it?

1. **What is the difference between an industry and a sector?**
2. **How many different industries are there?**

4.2. FACTORS AFFECTING LOCATION OF INDUSTRIES

Learning Objectives

- Understand the availability of raw materials and availability of labor
- Explain proximity to markets as well as transport facilities
- Focus on power, site and services
- Discuss about finance
- Analyze natural and climatic considerations
- Discuss personal factors
- Identify strategic considerations
- Focus on external economies and miscellaneous factors

The location of industries is generally determined by economic factors, though some industries may also be influenced by non-economic factors. The primary objective in their selection of specific locations for the placement of industries is profit maximization, which also suggests cost reduction. Let's examine the different elements that influence the location of industries.

Several factors pull the industry to a particular place and the factors affecting the location of industries are:

1. Availability of raw materials
2. Availability of Labor
3. Proximity to Markets
4. Transport Facilities
5. Power
6. Site and Services
7. Finance
8. Natural and Climatic Considerations

9. Personal Factors
10. Strategic Considerations
11. External Economies
12. Miscellaneous Factors

Let us learn the factors affecting the location of industries in detail below.

Keyword

The **labor cost index (LCI)** shows the short-term development of the labor cost, the total cost on an hourly basis of employing labor.

4.2.1. Availability of Raw Materials

The proximity of an industry to raw material sources is a critical factor in its location selection. The industry's cost of production would decrease if it were closer to the sources of raw materials. The majority of total costs in the majority of major industries are attributed to the cost of raw materials. As a result, the majority of forest and agro-based industries are situated close to the sources of their raw materials.

4.2.2. Availability of Labor

The industry requires a sufficient supply of inexpensive, skilled labor. The labor cost to total cost of production ratio, or what Weber refers to as the "Labor Cost Index," determines how drawn an industry is to labor centers. Initially, the cotton textile industry in the Bombay region was concentrated due to the availability of skilled workers in the interior parts of the region.

4.2.3. Proximity to Markets

Entrepreneurs must consider access to markets as a crucial element influencing the location of industries. Businesses producing large, bulky, or perishable goods are usually located near markets due to the impracticality of shipping over long distances. Industries such as bread and bakery, ice, tins, cans, and so on that are situated close to the markets may be able to lower their transportation expenses when distributing the final product. When it comes to industries producing consumer goods as opposed to producer goods, market accessibility matters more.

4.2.4. Transport Facilities

The location of industry is generally influenced by transportation facilities. Transportation in its three forms, namely, road, rail, and water all work together to play a significant role. Thus, the intersections of railroads, roads, and waterways develop into bustling hubs of industrial activity.

Furthermore, the location of industrial units is significantly impacted by the government's transport policy, rates, and modes of transportation. Bombay has a highly concentrated cotton textile industry because of its excellent and reasonably priced transportation network for both markets and raw materials.

Keyword

A **transport network**, or transportation network, is a network or graph in geographic space, describing an infrastructure that permits and constrains movement or flow.

4.2.5. Power

The availability of inexpensive power is another factor that affects where an industry is located. The main energy sources are coal, wind, water, gas, oil, and electricity. Prior to the development of the steam engine, many people looked to the wind and water for their energy needs.

When it came to the establishment of new industries during the nineteenth century, especially heavy industries, proximity to coal fields became the primary determining factor. Along with the development of alternative energy sources like oil, gas, and electricity, industry decentralization and dispersion resulted from the power factor's increased flexibility.

Let us dive further and learn some more factors affecting the location of industries.

Remember

Industrialization frequently leads to the expansion and growth of cities and towns.

4.2.6. Site and Services

The location of an industry is influenced to some degree by the availability of public utility services, the affordability of the site's value, and amenities specific to a given location such as the site's levelness, vegetation type, and proximity to related activities. The government has designated certain areas as underdeveloped, where entrepreneurs will receive various incentives such as subsidies, concessional financing, lower electricity rates, and access to facilities for education and training. These incentives may encourage some business owners to establish operations in these locations.

4.2.7. Finance

Finance is necessary for the establishment, operation, and growth of any industry. The availability of capital in sufficient quantities and at low interest rates is a major factor affecting the location of industries.

When the locational history of the Indian cotton textile industry is examined, for example, it becomes clear that wealthy and resourceful Parsi and Bhatia merchants played a major role in the industry's early concentration in and around Bombay by providing a large amount of funding.

Did you know?

The biggest information technology industry hubs are India's Bangalore region and Central California's Silicon Valley.

4.2.8 Natural and Climatic Considerations

Natural and climatic factors encompass various aspects such as land elevation, topography, water resources, waste management, and so on. Occasionally, these variables affect where industries are located.

For example, in the cotton textile industry, the *humid climate* is advantageous because it reduces the frequency of yarn breakage. Manchester, England, and Bombay, India, both have humid climates that provided excellent opportunities for the growth of cotton textile industries.

4.2.9. Personal Factors

An entrepreneur may occasionally have biases and personal preferences when choosing where to locate industrial units. For example, Mr. Ford began producing automobiles in Detroit simply because it was his hometown. In these situations, the individual's influence outweighs other factors. This kind of dominance is uncommon, though.

Keyword

The humid climate is the climate where precipitation is more than evaporation. It has fairly distributed precipitation throughout the year. There are three main humid climates: the humid subtropical climate, the humid tropical climate, and the humid continental climate.

4.2.10. Strategic Considerations

In the present day, industrial location decisions are significantly influenced by strategic factors. A secure location is becoming increasingly crucial during times of war. This is due to the fact that during armed conflicts, armament and ammunition factories, as well as businesses that provide other necessities for warfare, become primary targets of aerial strikes. An interesting example is Russia's experience during the Second World War.

4.2.11. External Economies

The location of industries is significantly impacted by external economies as well. When an industry is mostly concentrated in a single center with access to a port and shipping facilities, external economies are created by the expansion of specialized subsidiary activities. When several manufacturing facilities in the same industry are situated close to one another, external economies might also be realized.

4.2.12. Miscellaneous Factors

In some situations, historical events are also very important in deciding where industries are located. The growth of Lancashire's cotton textile industry offers an intriguing illustration of this. Additionally, a large factor in selecting a location for an industrial unit would be its size. This is so because the size of an industrial unit is determined by the radius of the circle in which it can distribute its goods profitably as well as by the population density inside the circle.

Question

What does a modern framework comprise?

Answer

The finished result and the pay acquired from it.

A modern framework comprises data sources, cycles, and results. The data sources are the unrefined components, work, and expenses of land, transport, power, and another framework. The cycles incorporate a wide scope of activities that convert the unrefined substance into finished products. Industries are groups of related businesses; department stores and shoemakers are two examples of industries. The primary product that a company produces or markets establishes the industry classification. Sectors are used to categorize industries. Similar businesses are categorized into different industries based on the main product that is produced or sold. This effectively creates industrial groups, which are then used to distinguish businesses from other entities that partake in different activities. Economists and investors regularly study industries to identify the variables that drive and constrain the expansion of corporate profits. A company's relative attractiveness within an industry can be ascertained by comparing it to other companies in that same industry.

Did you know?

Eastern Asia, Western Europe, Central Europe, Eastern Europe, and Eastern North America are the world's major industrial regions.

Did you get it?

1. **Depict momentarily the arrangement of ventures on the premise of unrefined substance used.**
2. **What are the data sources, cycles, and results of a material industry?**

4.3. INDUSTRIAL DEVELOPMENT AND ECONOMIC GROWTH

Learning Objectives

- Understand the high-growth industries
- Identify characteristics of the growing industry
- Discuss about economic growth
- Explain industrialization and growth
- Analyze the progression of the Industrial Revolution
- Focus on the role of industrial development in economic growth
- Identify challenges and opportunities in modern industry
- Understand the role of policy and regulation in the industry
- Know the future of the industry

Industrialization has long been recognized as a crucial factor for regional economic development. It plays a significant role in the secondary sector, aiming to increase the region's per capita income and offer employment opportunities to the growing population to meet the demand for manufactured goods. Industry is responsible for converting raw materials into finished products, which in turn supports the growth of the tertiary sector by providing services, transportation, and marketing, among other things. Industrialization is often associated with social changes, such as the shift of people from the primary to secondary and tertiary sectors.

4.3.1. Understanding the High-Growth Industries

The development of a booming industry is influenced by numerous variables. New technologies can inspire startups and entrepreneurs to develop new products and services related to the industry. One example of this is the introduction of creative and innovative technologies. With technology always evolving, investors are lured to invest in the technologies because they offer the prospect of exponential growth in the future. The smartphone business, which merged multiple groundbreaking technologies into a single mobile device, expanded quickly in the first half of the current decade. Three examples of this approach that are used today are virtual reality (VR), artificial intelligence, and *machine learning*. An entirely virtual, computer-generated scenario that can be used to mimic a real-world setting is called a virtual reality (VR) experience. Applications for it include driving simulators, medical school education, and gaming headsets.

Keyword

Machine learning is a branch of artificial intelligence (AI) and computer science that focuses on the use of data and algorithms to imitate the way that humans learn, gradually improving its accuracy.

4.3.2. Characteristics of the Growing Industry

Businesses in growth sectors tend to experience rapid and steady increases in sales figures along with a capital inflow. Typically, unwelcome media attention follows this. High-growth industry stocks are riskier and more erratic. Frequently, investors are willing to assume supplementary risks in return for the potential to earn substantial returns.

High capital burn rates, profitability gaps despite investor and consumer enthusiasm, bubbles, and technological setbacks that impede advancement are among the risks associated with developing sectors.

4.3.3. Economic Growth

There are a few recognized strategies for achieving substantial economic growth. The first is trade specialization, which enhances a laborer's performance through knowledge, experience, and education. Specialization usually occurs organically as individuals and companies seek to maximize their profits from trade.

The second known strategy is utilizing better capital goods. Productivity per labor hour increases with improved tools. For example, a person riding a bicycle and carrying a backpack cannot transport goods over a distance as efficiently as an 18-wheeler.

The final strategy for enhancing productivity is to identify previously underutilized resources. The development of the Internet and oil wells in the 1850s are two examples of this approach.

More rapid production of goods results in a decrease in acquisition costs. It is now simpler for individuals and families to buy these goods thanks to declining real costs. As a result, living standards are raised. Without increases in productivity, most families would be unable to afford to own TVs, computers, refrigerators, cars, running water, electricity, or any number of other goods.

4.3.4. Industrialization and Growth

According to estimates from the Federal Reserve branch in Minneapolis, human productivity and corresponding standards of living were essentially unchanged from the beginning of the agricultural age until 1750 A.D.

In Great Britain, things began to change in the 1760s. The population and average income started rising at a rate never seen before. GDP per capita had been constant for thousands of years, but with the advent of the modern capitalist economy, it experienced a sharp increase.

Economic historian Deirdre McCloskey, writing in the Cambridge University Press in 2004, argued that industrialization was "certainly the most important event in the history of humanity since the domestication of animals and plants, perhaps the most important since the invention of language."

Regarding what lit the Industrial Revolution, historians cannot all agree. The shifts in Great Britain's legal and cultural framework that permitted free trade and provided entrepreneurs with the space and motivation to take chances, innovate, and make money are often cited by economists.

4.3.5. Progression of the Industrial Revolution

Through the invention of more advanced capital goods, like the steam engine, and the mastery of production methods, like the assembly line, marginal productivity increased significantly during this time.

Products were made in comparatively shorter amounts of time. Both population growth and the fight against malnutrition were aided by more and better food. Education, creativity, and leisure were given more time. The sharp increase in average real incomes fed the demand for better products and services.

Question

What does agrarian mean?

Answer

Agrarian is a term that relates to farms and farming but its scope is broader than that. It also encompasses issues of land and family ownership of that land in politics and society. Agrarianism represents a segment of the economy and is believed to promote a healthier, more stable lifestyle.

4.3.6. Role of Industrial Development in Economic Growth

The following points explain the role of industrial development in economic growth.

4.3.6.1. Modernization of Industry

Agriculture cannot be modernized without industrial development. Agriculture in India is archaic and traditional. Productivity is low and the cost of production is high. To modernize agriculture, we need harvesters, tractors, threshers, and pump sets. Chemical weed killers, insecticides, and fertilizers are necessary to boost productivity. All of these items are manufactured. These goods can't be produced without industrial development. Goods derived from agriculture, such as sugarcane, cotton, and jute are unprocessed substances. To prepare final goods such as sugar, flex, and textiles, among others we must become more industrialized. Thus, the modernization of agriculture requires industrial development.

Keyword

A **by-product** or byproduct is a secondary product derived from a production process, manufacturing process or chemical reaction; it is not the primary product or service being produced.

4.3.6.2. Development of Science and Technology

Science and technology advance in tandem with industrial development. Businesses in the industrial sector carry out research and create new goods. One example of industrial development is the production of biofuel from ethanol. The industry studies its wastes and creates byproducts, such as biodiesel from Jatropha seeds. Industrialization has allowed us to advance in fields like atomic science, satellite communication, missile technology, etc.

4.3.6.3. Capital Formation

The primary issue facing the Indian economy is a severe lack of capital. The surplus in the agricultural sector is minimal. It is also quite challenging to mobilize. The surplus is extremely high in large-scale industries. Industry is able to increase profit by utilizing both internal and external economies. These gains may be put back into growth and development projects. Industrialization thus aids in the creation of capital.

4.3.6.4. Industrialization and Urbanization

Industrialization gives way to urbanization. A region's transportation and communication systems expand as a result of industrialization. Educational institutions, medical centers, technical institutes, banks, and schools are located in close proximity to an industrial base. Rourkela, an ultramodern town in Odisha, was once covered in dense forest. Following the establishment of a large industry, numerous auxiliary units were created.

4.3.6.5. Self-Reliance in Defense Production

Industrialization is necessary to achieve self-reliance in defense production. Depending on other nations for military equipment during emergencies or conflicts could be dangerous. It is also crucial to have a self-sustaining industrial infrastructure and capital goods. The Pokhran (Rajasthan) atomic explosion and the Agni Missile are two examples of industrial expansion.

Keyword

Industrialization is the process of transforming the economy of a nation or region from a focus on agriculture to a reliance on manufacturing.

4.3.6.6. Importance in International Trade

Industrialization significantly supports trade promotion. Trade benefits more developed countries compared to less developed ones. Developing nations export primary goods and import industrial goods. Generally, agricultural products have elastic demand and lower prices, while industrial products have a stable demand and higher values. This leads to trade imbalances. To address the balance of payments deficit, we need to either produce goods to replace imports or focus on export promotion through industrial development.

4.3.6.7. Use of Natural Resources

India is often seen as a country rich in resources but with many people living in poverty. This suggests that despite having abundant natural resources, India hasn't fully utilized them due to limited funding and technology. It's important to use resources wisely when turning them into industrial products. In the past, India provided the British with affordable raw materials for their manufacturing, while also serving as a market for their goods. This helped England during the Industrial Revolution, but India continued to face poverty. Therefore, *industrialization* plays a key role in efficiently utilizing available resources.

4.3.6.8. Alleviation of Poverty and Unemployment

With fast industrialization, poverty and unemployment can be eliminated. It has happened in developed nations with advanced industries, like Japan. Widespread poverty and high unemployment are caused by the industrial sector's sluggish growth. Therefore, villages with excess labor can use it in industry due to the rapid growth of the industrial sector.

4.3.6.9. Main Sector of Economic Development

A key driver of economic development is thought to be industry. By using sophisticated technology, labor division, and scientific management, we can achieve economies of scale. Thus, there will be a sharp rise in output and jobs. Economic expansion and capital formation will result from this.

4.3.6.10. Fast Growth of National and Per Capita Income

Industrial development helps in the rapid growth of national and per capita income. The history of economic development of advanced countries shows that there is a close relation between the level of industrial development and the level of national and per capita income. For instance, the share of the industrial sector in national income was 26% and the per capita income in the year 2000 was $36,240 in the USA. The share of agriculture in the same year was only 2%. In Japan, the share of the industrial sector in her GDP was 36% and her per capita income was $36,210.

Keyword

Digital technologies are electronic tools, devices, systems, and resources that generate, store or process data.

4.3.6.11. Sign of Higher Standard of Living and Social Change

A country cannot produce goods and services of high quality in order to attain decent living standards without the progress of the industrial sector.

The world's economic health—or lack thereof—is largely due to industrialization. For centuries, industrialization has led to significant improvements in production, which have made people's lives much nicer and more comfortable.

4.3.7. Challenges and Opportunities in Modern Industry

Modern industries face opportunities as well as challenges in navigating the complexities of globalization, technological advancements, and sustainable growth. To succeed and stay competitive, one must adjust to this quickly shifting environment.

Industries must take advantage of opportunities and overcome numerous challenges to survive in the modern world. Now let's take a closer look at some of the major factors that influence the industrial landscape.

4.3.7.1. Navigating Globalization and Technological Changes in Industry

Industries need to figure out how to handle the opportunities and problems that come with globalization. Industries must be flexible and agile in response to shifting consumer demands, global supply chains, and heightened competition.

New markets and increased chances for industries to connect with a larger consumer base have been brought about by globalization. On the other hand, as businesses from all over the world compete for market share, it has also increased competition. Industries need to continuously innovate and find methods to set themselves apart if they want to stay ahead of the curve.

Technological developments are still upending industries and changing their environments. Whether it is through the increasing use of automation, e-commerce, or the adoption of *digital technologies*, industries need to welcome innovation and always look for new ways to use technology to their advantage.

Remember

By embracing change, investing in their workforce, and fostering a culture of innovation, industries can position themselves for success in the ever-evolving industrial landscape.

For instance, the emergence of e-commerce has completely transformed the retail sector by enabling companies to connect with clients online as well as offline. Manufacturing procedures have been simplified by automation, increasing productivity and cutting expenses. Industries must adopt these technological advancements if they want to compete in the global marketplace.

4.3.7.2. Environmental Sustainability and Industry: Balancing Growth with Conservation

Environmental preservation and the goal of economic expansion must coexist. The global effort to combat climate change, lower emissions, and safeguard natural resources is heavily reliant on the industrial sector. Industries can reduce their adverse effects on the environment by implementing sustainable practices, such as using waste reduction techniques, efficient production processes, and the adoption of renewable energy sources. Sustainable business practices improve brand reputation, help businesses succeed overall, and are good for the environment.

For example, the automotive sector has achieved great progress in the development of electric cars and the reduction of carbon emissions. Sustainable materials and ethical sourcing methods are being investigated by the fashion industry. These initiatives appeal to environmentally conscious consumers in addition to helping to create a greener future.

4.3.7.3. Addressing the Skill Gap and Workforce Challenges in Industrial Sectors

An increasingly skilled and flexible workforce is required as industries adopt new technologies and go through digital transformations. Nonetheless, a lot of industries struggle to find employees who have the skills and knowledge needed to succeed in the contemporary industrial environment. Governments, businesses, and educational institutions must work together to close the skill gap. To close the skills gap and guarantee a workforce prepared to handle shifting industry demands, creative training programs, apprenticeships, and continuing education initiatives can be implemented.

For industries to stay up to date with technological advancements, they must also invest in reskilling and upskilling their current workforce. Industries may enable their workers to adapt and prosper in the quickly changing industrial landscape by offering opportunities for professional development and ongoing education.

4.3.7.4. The Role of Innovation and R&D in Industrial Competitiveness

Research and development (R&D) along with innovation are key factors that propel industrial competitiveness. Businesses that make innovation investments cultivate a culture of ongoing improvement and are better able to adjust to shifting consumer preferences and market conditions. Government grants and incentives for research and development can greatly increase industry innovation. Industry-research institution collaborations can yield ground-breaking discoveries and technological innovations that propel industries forward to new heights.

For example, the pharmaceutical industry heavily relies on R&D to develop new drugs and treatments. The technology sector thrives on innovation, constantly pushing the boundaries of what is possible. Embracing a culture of innovation and investing in R&D is crucial for industries to stay ahead of the curve and maintain their competitive edge.

As industries navigate the challenges and opportunities in the modern world, they must remain adaptable, innovative, and conscious of their impact on the environment.

4.3.8. The Role of Policy and Regulation in Industry

Policies and rules governing the industry shape the industrial environment and control industry operations. Ensuring fair competition and promoting industrial growth are largely dependent on trade agreements, international standards, and government policies.

Government policies can foster an environment that is conducive to industry growth. Innovation and the expansion of industry can be stimulated by policies that support investment, offer financial incentives, and aid in research and development. To encourage businesses to invest in new technologies or expand their operations, governments could, for instance, provide tax breaks or grants. The industries themselves gain from these policies, but they also advance general economic growth.

Government policy has the potential to stimulate industrial diversification in addition to investment. Governments can lessen their reliance on particular industries and improve economic resilience by assisting businesses in a variety of sectors. Countries that are more diversified are better able to withstand economic downturns and adjust to shifting global market trends. In addition, policies aimed at generating jobs can also significantly lower unemployment rates and enhance the general welfare of the populace.

4.3.8.1. Government Policies Shaping the Industrial Landscape

The industrial landscape is significantly shaped by government policies. They possess the ability to impact the course and expansion of industries within a nation. By implementing policies that align with national development goals, governments can foster the growth of specific industries that are deemed strategically important.

Keyword

A **waste management system** is a streamlined process that organizations use to dispose of, reduce, reuse, and prevent waste.

For example, a government may designate the development and adoption of clean energy technologies as a priority and enact policies in support of this goal. These regulations may call for a specific percentage of energy to come from renewable sources, or they may include tax breaks for businesses that invest in renewable energy projects and subsidies for renewable energy projects.

Policies from the government can also address environmental issues and encourage industry-wide sustainable practices. The detrimental effects of industrial activity on the environment can be lessened by regulations requiring businesses to use environmentally friendly production techniques, reduce their carbon emissions, or put *waste management systems* in place.

4.3.8.2. International Standards and Compliance in Industrial Operations

Industries that compete in a globalized market must adhere to international standards. International standards protect consumers and foster fair competition by ensuring that goods and services fulfill quality, safety, and environmental requirements.

In order to attain global market access and establish a reputation for dependability and excellence, industries are required to conform to international certifications and regulations. For instance, ISO certifications are globally accepted and serve as evidence of a business's dedication to fulfilling global requirements in a number of domains, including environmental management, occupational health and safety, and quality management. In addition to helping industries, adhering to international standards levels the playing field for companies that operate internationally. It guarantees that businesses everywhere are held to the same standards and aids in stopping unfair competition practices.

4.3.8.3. The Importance of Trade Agreements and Economic Alliances

Economic alliances and trade agreements open up new markets for businesses, encourage foreign direct investment, and improve collaboration across borders. These agreements encourage business opportunities, industrial growth, and economic integration.

Economic growth and innovation can be stimulated by industries expanding their reach and accessing new consumer markets through the removal of tariffs, trade barriers, and harmonization of standards. Trade agreements offer a structure for settling conflicts and safeguarding intellectual property rights, which are crucial for sectors of the economy that depend on technological advancement and innovation.

Economic alliances can help industries in different nations collaborate and share knowledge. This knowledge and idea sharing can result in the creation of new technologies, enhanced manufacturing procedures, and heightened competitiveness.

4.3.8.4. Regulatory Challenges and Compliance in Industrial Sectors

Standards and regulations are necessary for industries to run smoothly, but they can also be problematic. Industries have a lot of regulations to follow, and they have to navigate complicated regulatory frameworks. Regulatory burdens can impede innovation and growth while adding to industry costs. Governments and regulatory agencies must therefore strike a balance between maintaining an environment that promotes industry development and innovation and enforcing the laws that are necessary. Maintaining this equilibrium necessitates constant communication between legislators and industry representatives to guarantee that rules are practical, efficient, and responsive to business requirements. Regulation compliance frequently necessitates large infrastructure, technology, and human resource investments. Small and medium-sized businesses' (SMEs') limited resources may make it especially difficult for them to meet these requirements. As a result, governments can be extremely helpful in supporting and assisting SMEs so they can adhere to rules and compete on an even playing field with bigger businesses. Regulation and policy are essential in determining the industrial landscape. While international standards and trade agreements encourage fair competition and access to global markets, government policies can also foster an environment that is conducive to the growth of industries. To ensure that industries can develop and grow while satisfying societal and environmental needs, it is crucial to strike the correct balance between regulation and industry development.

4.3.9. The Future of Industry: Trends and Predictions

Anticipating future trends and formulating strategies to capitalize on emerging opportunities are critical as industries undergo continuous evolution. The future of industries and the path for the subsequent stage of economic growth and innovation will be shaped by Industry 4.0, emerging technologies, and sustainable practices.

By embracing *digital transformation* and investigating new opportunities, industries need to stay ahead of the curve. They will be able to take advantage of the full potential of cutting-edge technologies and gain a market advantage by doing this.

4.3.9.1. Anticipating the Future of Industry in a Digital World

The digital revolution has completely changed how industries function. In order to increase productivity, boost efficiency, and develop new business models, industries will continue to adopt digital technologies, automation, and data analytics in the future. Envision an environment where machines coexist harmoniously with humans, utilizing artificial intelligence and machine learning techniques to enhance workflows and generate self-governing choices. In addition to streamlining operations, this degree of automation will allow industries to deliver goods and services with never-before-seen accuracy and speed.

Keyword

Digital transformation is the incorporation of computer-based technologies into an organization's products, processes and strategies.

4.3.9.2. The Role of Emerging Technologies in Transforming Industries

Emerging technologies hold incredible potential to revolutionize industries across the spectrum. From renewable energy sources and biotechnology to advanced materials and nanotechnology, these technologies will reshape industries in ways we have yet to fully comprehend.

Industries must embrace and invest in emerging technologies to stay competitive and drive innovation. Collaborations between industries, research institutions, and startups can accelerate the adoption and development of these technologies, unlocking new possibilities for industrial growth.

Imagine a future where renewable energy sources power entire manufacturing processes, reducing carbon emissions and creating a sustainable future. This shift towards clean energy will not only benefit the environment but also open up new avenues for economic growth and job creation.

4.3.9.3. The Potential of Industry 4.0 and Smart Manufacturing

Industry 4.0 is a paradigm shift in the industrial landscape that combines automation, connectivity, and digital technologies. Industry 4.0's smart manufacturing technology will give rise to intelligent factories, where machines can interact with one another and decide for themselves.

With the power of real-time data analytics, industries will be able to optimize production processes, identify inefficiencies, and respond swiftly to changing market demands. The potential for cost savings, enhanced productivity, and value creation through Industry 4.0 is immense.

Envision a factory where devices are networked together, sharing data continuously and modifying their functions in response to real-time information. This degree of intelligence and connectedness will transform manufacturing, resulting in higher *productivity*, less downtime, and better-quality products.

4.3.9.4. Preparing for the Challenges of Automation and AI in Industry

There are increasing benefits and drawbacks to industries using automation and artificial intelligence (AI). Even though these technologies can improve efficiency and productivity, there are worries about job displacement and the impact on the labor force as well. Businesses need to invest in worker reskilling and upskilling initiatives in order to proactively address these issues. In order to facilitate a seamless transition and generate new opportunities in fields that emerge from the convergence of technology and industry, governments, industries, and educational institutions must collaborate.

Imagine a time when laborers have the know-how to survive in an automated society. By offering extensive training programs and cultivating an environment that values ongoing education, industries can enable their workforce to adopt new technologies and adjust to a dynamic environment. The industrial sector has enormous growth and innovation potential in the future. Industries can take the lead in the next industrial revolution by embracing digital transformation, making investments in cutting-edge technologies, and tackling the issues posed by automation and artificial intelligence. To build a sustainable and prosperous future for everybody, the secret is to foresee future trends, formulate strategies, and promote teamwork.

Industry is undeniably the backbone of economic growth and innovation. Its evolution and influence on the global economy have been transformative, shaping nations and propelling socio-economic progress. As industries face challenges and embrace new opportunities, collaboration and innovation will be key to unlocking their full potential. Policymakers, industries, and individuals must work together to navigate the complexities of globalization, technology advancements, and sustainability while fostering inclusivity and economic stability. We can create the conditions for a future in which industries remain the main engines of innovation and economic growth by exercising foresight, making strategic plans, and committing to equitable and sustainable industrial development.

Keyword

Productivity is commonly defined as a ratio between the output volume and the volume of inputs.

Did you get it?

1. **What is marginal productivity?**
2. **What is gross domestic product and why is it so important?**

FOCUS ON CAREERS

INDUSTRIAL ENGINEER

Industrial Engineers are responsible for planning, organizing, supervising, and managing the industry's operations to ensure economical, safe, and effective usage of materials, machines, information, people, and energy. They devise efficient systems that are essential for the manufacturing of products or services.

They are the problem-solvers who put on their creative thinking hats and knowledge of manufacturing technology to build strong organizational efficiency. These engineers always find ways to eliminate processing and production waste and make the best use of workers, time, money, and energy.

Responsibilities

- Review of production schedules, engineering specifications, and process flows to gain an understanding of applied methods and activities in manufacturing and services.
- Manufacture machine parts and products, and deliver services with maximum efficiency by using new, existing, or modified machinery or equipment.
- Conduct efficient manufacturing and cost analysis by designing and developing management control systems.
- Resolve problems related to production and cost minimization by implementing quality control procedures.
- Ensure products meet quality standards by designing production systems responsible for product planning and activity coordination.
- Host client meetings for product specifications, consult with vendors on purchases, and discuss manufacturing capabilities and project status with manufacturing personnel.
- Introduce or recommend changes to safety measures, work methods, and labor organization.

Education and Eligibility

- Complete 10+2 in the MPC stream with a minimum of 50% from a recognized board.
- Qualify for entrance exams like EAMCET, JEE Mains, or JEE Advanced.
- Pursue undergraduate courses in industrial engineering, electrical engineering, mechanical engineering, process engineering, and manufacturing engineering.

- Pursue a Master's in Industrial Engineering after graduation and qualifying in the GATE exam.
- Interested candidates can pursue a Doctoral degree or PhD.

Required Skills

- Excellent communication skills.
- Problem-solving and analytical skills.
- Computer Aided Design (CAD).
- Critical thinking skills and creativity.
- Technical writing skills.
- Excellent computer literacy.
- Knowledge of factors influencing industrial efficiency.
- Interpersonal and negotiation skills.
- Basic finance skills.
- Advanced algebra, geometry, calculus and statistics.
- Business Strategy, operations and documentation skills.

Specializations in Industrial Engineering

- Operations research
- Defense-oriented operations research
- Production and management
- Manufacturing
- Health systems
- Financial engineering
- Human factors
- Risk management
- Supply chain management

Salary and Job Prospects

The median annual salary for Industrial Engineers is $84,310. In India, these engineers earn an average of INR 4–5 LPA yearly as entry-level engineers. This salary is dependent on factors like education level, experience, location, and industry. According to the U.S. Bureau of Labor Statistics, the field of Industrial Engineering is expected to grow at a rate of 8% from 2018–2028.

These engineers can be employed in various industries like major manufacturing industries, engineering and consultant services, wholesale trade, research and development. Also, the increasing growth in the healthcare industry and how it is delivered is also responsible for creating demand for industrial engineers in professional, scientific, and consulting firms.

Entry-level engineers may work under the supervision of experienced industrial engineers with formal training in large companies. With experience and skills, they may advance to technical specialist roles like quality engineers and facility planners where they lead a team of engineers or technicians. They may also move into management roles as they mostly perform the work of managers. Pursuit of a master's degree is responsible for such specializations and advancement.

A CLOSER LOOK

INDUSTRIALIZATION

Industrialization can be driven by a combination of factors including government policy, labor-saving inventions, entrepreneurial ambitions, and a demand for goods and services. It has profound implications for the population, causing a wave of migration from small farms to cities and towns where jobs can be found.

The most dramatic example in recent history is that of China, where government policy changes in the late 20th century led to the nation's transition from an economy based on subsistence farming to a global manufacturing powerhouse.

The Industrial Revolution

In the western world, Industrialization is most commonly associated with the Industrial Revolution in Europe that began in the late 18th century and the subsequent burst of industrialization in the U.S. through the 19th century.

In Europe, the era was characterized by a surge in local manufacturing of goods for export, made possible by a growing population of consumers. Great Britain played an outsized role in the process through technological innovations such as steam-powered machinery. Industrialization quickly spread to the United States, the epicenter of laissez-faire capitalism. Inventions including the cotton mill and steam power made possible the establishment of mill towns such as Lowell, MA, and Pawtucket, RI.

Later Periods of Industrialization

World War II created an unprecedented demand for certain manufactured goods, leading to a buildup of production capacity. Post-war prosperity provided further catalysts that kept capacity utilization high and stimulated further growth.

Innovation, specialization, and wealth creation were the causes and effects of industrialization in this period.

The Asian Tigers

The late 20th century was marked by rapid industrialization in other parts of the world, notably Asia. The Asian Tigers (Hong Kong, South Korea, Taiwan, and Singapore) all participated in economic growth based on manufacturing for global customers.

China experienced its own industrial revolution after moving away from a strict communist model.

SUMMARY

- The main product that similar businesses manufacture or sell is what categorizes them into industries. This gives rise to industry associations, which subsequently serve as a means of severing companies from various other groups.
- The amount of money invested, the goods produced, and the number of employees involved in this case are used to measure the size of an industry. Industries can be classified as either large or small.
- Systems of classification help group related business operations into sectors and industries. Sectors tend to encompass a broader scope than industries.
- The proximity to raw material sources is crucial in selecting a location for an industry. Being closer to these sources would reduce the industry's production costs.
- Additionally, the industry needs an ample supply of affordable, skilled labor.
- One of the crucial elements influencing the location of industries that entrepreneurs must take into account is access to markets. Businesses that produce large, bulky, or perishable goods that are impractical to ship over great distances are typically found near markets.
- The location of industries is mainly influenced by transportation facilities. Road, rail, and water transportation all collaborate to play a crucial role.
- The availability of affordable power is another factor that influences the location of an industry. The primary energy sources include coal, wind, water, gas, oil, and electricity. Before the steam engine was developed, many relied on wind and water for their energy requirements.
- The location of an industry is influenced by public utility services, site value, and specific amenities like levelness, vegetation, and proximity to related activities.
- Industrialization is crucial for regional economic development.

Industry growth is driven by various factors, including new technologies inspiring startups to create new products and services.

- Businesses in growing sectors experience rapid sales increases and capital inflow.
- Producing goods more quickly leads to lower acquisition costs, making it easier for people and families to purchase these items due to reduced real costs.
- During this period, advancements in capital goods, such as the steam engine, and improvements in production methods, like the assembly line, led to a significant increase in marginal productivity.
- Modern industries must adapt swiftly to the challenges and opportunities presented by globalization, technological progress, and sustainable growth to maintain competitiveness.

MULTIPLE CHOICE QUESTIONS

1. **The Industries are classified on the basis of?**
   a. Size
   b. Ownership
   c. Raw materials
   d. All of the above
2. **How do we measure the size of the industry?**
   a. Money invested
   b. Labor force
   c. Advance technology
   d. All of the above
3. **Which of this is a joint sector industry?**
   a. Only owned by individuals
   b. Owned by the government
   c. Co-Owned by the government and individuals
   d. Owned by producers, workers, and suppliers.
4. **___________facilities, generally, influence the location of industry.**
   a. External economics
   b. Temperature
   c. Personal factors
   d. Transport
5. **Existence of public utility services, the cheapness of the value of the site, amenities attached to a particular site like ____________.**
   a. Level of air
   b. Level of atmosphere
   c. Level of temperature
   d. Level of ground
6. **Most of the ________ and forest-based industries are located in the vicinity of the sources of raw material supply.**
   a. Cotton textile
   b. Crude oil
   c. Paper manufacturing
   d. Agro-based
7. **__________ producing perishable or bulky commodities that cannot be transported over long distances are generally located in close proximity to markets.**
   a. Machines

b. Companies

c. Small scale

d. Industries

8. **The value of goods and services produced by residents of a country and the value of their property is called_________.**

a. Gross domestic product

b. Net domestic product

c. Gross national income

d. Net national income

9. **Gross domestic product of a country is the total of its net domestic product and ________.**

a. Informal economy

b. Earnings on foreign exchange in the domestic country

c. Subsidies on the consumption of goods and services

d. Consumption of capital in the production process

10. **The meaning of disguised unemployment is _________.**

a. Zero marginal productivity of labor

b. Zero total productivity of labor

c. Zero average productivity of labor

d. None of the above

REVIEW QUESTIONS

1. What is the main basis for industry?
2. What are the main factors which influence the location of an industry?
3. Name any three physical factors for the location of the industry.
4. Name any two human factors that affect the location of industries.
5. How did the Industrial Revolution change economies?
6. Where and when did the Industrial Revolution take place?
7. How did the Industrial Revolution change society?
8. Who were some important inventors of the Industrial Revolution?

Answer to Multiple Choice Questions

1. (d)	2. (d)	3. (c)	4. (d)	5. (d)
6. (d)	7. (d)	8. (c)	9. (d)	10. (a)

REFERENCES

1. Feldstein, M., (2008). *The United States in the World Economy.*
2. Garay, R. G., (2011). *U.S. Steel and Gary, West Virginia.* Univ. of Tennessee Press.
3. Jomo, K. S., (2001). *Southeast Asia's Industrialization*. London: Palgrave Macmillan UK. http://dx.doi.org/10.1057/9781137002310.
4. Nawratek, K., (2017). *Urban Re-industrialization*. Earth, Milky Way: Punctum Books.
5. Sandten, C., Gunter, S., & Evelyne, K., (2009). *Industrialization, Industrial Heritage, De-industrialization*. Trier: WVT, Wissenschaftlicher Verlag Trier.
6. Shoven, J. B., (2006). *Government Policy Towards Industry in the United States and Japan.* Cambridge University Press.

CHAPTER 5

Transport and Communication

LEARNING OBJECTIVES

After studying this chapter, you will be able to:

- Understand land transport
- Discuss water transport
- Focus on air transport
- Identify major issues in transport and communication

"We must accept the fact that transport and communications will bring the world into close relations, and the youth of the world should have standards and ideals in common."

—Juliette Gordon Low.

INTRODUCTORY EXAMPLE

History of Transportation

Beginning with the emergence of humans, transportation has undergone constant change throughout history. The human foot was the first means of propulsion. Long walks were once required to get places. Adapting to varied surfaces was the first improvement made to this mode of transportation. For instance, in order to prevent slipping on the ground, residents of places with snow and ice wore attachments that resembled spikes.

People knew that trees and logs float on water, so they dug out the middle portion of logs to form a kind of seating. This helped people use water bodies as a means of transport. The earliest vehicles with wheels were used around 3500 BC. Wheels were added to carts and chariots to facilitate the transportation of small loads. People turned ordinary logs into maneuverable riverboats with oars to steer the vehicles around the same period that transportation history was created. After that, people began training animals for use as a means of transportation, such as horses. Then, domesticating animals to use them to transport people and small goods began to catch on.

Change in Transportation History with the Discovery of the Wheel

The invention of wheels significantly altered the history of transportation. Other smaller devices like wheelbarrows were used after the wheel and axle were discovered. After that, existing modes of transportation were constantly enhanced. For instance, it became customary to use iron horseshoes. There was then a chance to combine various forms of transportation. Horse-drawn carriages and carts are one example. From this point on, transportation-related advancements began to pick up significant speed. Submarines first appeared around 1620, and by the 1660s, there were functional public transportation options. Large-scale operational vehicles were developed from carriages, steamboats, bicycles, and even hot air balloons. In 1867, the first motorcycle was created, following the invention of the first gas-powered vehicle by Jean Lenoir in 1862. Ultimately, the Wright brothers created the first powered human-piloted aircraft in 1903. The first successfully launched liquid-propelled rocket occurred in 1926, and subsequent vehicles such as helicopters, jets, and hovercrafts were developed. Transportation options already in place were always being enhanced. Trains with bullet trains were made possible by steam engines. From walking to flying, we have come a long way, and various modes of transportation must connect to a vast network. The Wright brothers' invention of manned flight led to the development of the Jumbo Jet. The types of transportation in existence now are given below.

- Land Transport
- Railways
- Water Transport
- Air Transport or Aviation
- Space Travel

INTRODUCTION

Transport moves goods from the sites of production to the sites of consumption, facilitating trade and commerce. Items that are lacking in certain areas are supplied with goods from the excess areas. People move around using transportation facilities in search of employment, education, or emergencies. We remain informed about global events and trends thanks to communication. It improves people's lives in positive ways, which boosts their economic circumstances.

5.1. LAND TRANSPORT

Learning Objectives

- Understand the history of land transportation
- Identify users of land transportation
- Explain the advantages and disadvantages of land transportation
- Describe the impact of land transportation

Land transportation, in various forms, facilitates the movement of people, animals, goods, and services globally. Initially, walking was the primary mode of transportation, but options expanded over time. Land transportation encompasses pipelines, tunnels, roads, railroads, bridges, and seaports. Railroads and roads are now the most popular forms of land transportation, making travel easier for all.

5.1.1. History of Land Transportation

With the long history of land transportation, the definition of transportation has evolved. Walking was the main mode of transportation to get to any location, no matter how far or close, as was already mentioned. Regardless of the distance and difficulty, humans and animals have long used this method.

To transport passengers, manufactured goods, food, and raw materials, strong animals were employed. Animals saved human labor and effort by covering the same distance in half the time. The invention of wheels and carts increased accessibility for people who wanted to transport goods.

The ability to transport people and goods over roads has been greatly enhanced. The introduction of *steam engines* gave land transportation a new perspective. In contrast to vehicles powered by animals, steam engines were roomy, cozy, and quick. Engines have made communication quick and simple while also promoting independence.

Subsequently, railroads gained prominence as a means of land transportation after the introduction of motor engines. For the community, railroads have created a large

Did you know?

The world railway network as a whole developed back at the beginning of the 20th century, its length now amounts to 13.2 million km with considerable unevenness of location.

number of job opportunities. Over time, buses and private vehicles became more widespread. As land transportation options have expanded over time, people's lives have become quicker and easier.

5.1.2. Users of Land Transportation

The two main categories of land transportation are railroads and roads. Among the various modes, road transportation is the oldest. Walking, cycling, animal carts, driving a car, etc., are examples of road transportation. Airports and seaports can be reached via road transportation as well as other forms of transportation (Figure 5.1).

Figure 5.1. Intercity Express, a German high-speed passenger train.

Source: By S. Terfloth – Own work, CC BY-SA 2.0 de, https://commons.wikimedia.org/w/index.php?curid=1666910

Keyword

A **steam engine** is a heat engine that performs mechanical work using steam as its working fluid.

The purpose of railroads was to transport heavy cargo, such as manufactured goods, fuel, food, and raw materials. Diesel-powered motor engines eventually replaced steam engines as the primary form of propulsion. Trains can now be powered by gas turbines or electricity. The general public uses rail and road transportation daily to fulfill their needs. They are also employed in commerce since land transportation makes it easier to move products and services.

5.1.3. Advantages and Disadvantages of Land Transportation

For hundreds of years, land transportation has been used by people all over the world. The benefits and drawbacks of land transportation are listed below.

Advantages:

- It connects the rural population to the urban population and erases the gap. Land transportation contributes to the development of rural areas and corporate change in lifestyle.
- Land transportation serves door-to-door reach of passengers and services both, irrespective of the distance.
- It is a capital-intensive mode of land transportation in comparison to water, railways, and air transportation.
- Land transportation is the priority for a short distance destination, as it doesn't require fast service.
- It offers safety to fragile items supplied to long or short distances.
- Motor mode of transportation provides aid to the rapid speed and safe times and is cost-effective.
- The various modes of vehicles like buses, trucks, minivans are used for public transportation by the government.
- The privilege of a private vehicle is used via land transportation for its comfort and luxury.

Disadvantages:

- Damaged roads can slow down the movement, and seasonal reasons like floods and natural disasters may be a barrier.
- The improper management & condition of land transportation may cause accidents and mishaps.
- Heavy traffic can increase travel time and may cause a delay in arrival at the destination.
- Many times the unorganized schedule of railways causes delay and time-killing.

Remember

Road transport involves the movement of goods or people by road, using various types of vehicles such as cars, trucks, and buses.

Question

Which transport is connected with land transportation?

Answer

Land transportation connects to seaports and airline transportation.

5.1.4. Impact of Land Transportation

The effects of land transportation are evident in people's lives everywhere. However, there are advantages and disadvantages to land transportation. These two aspects are as follows.

- Economy

The entire economy benefits greatly from land transportation. A rapid shift in the global market structure has been observed due to the streamlined movement of passengers, goods, and services. Trade between nearby countries and even those further away has been made possible by land transportation. It is affordable and practical.

Keyword

Petroleum is a fossil fuel that is used widely in the daily lives of global consumers.

- Environment

Over time, the innovation has gained appreciation as it has made traveling more comfortable and easy for travelers. The excessive use of non-renewable resources like coal and petroleum has an impact. Motor engines or vehicles run on petroleum, but steam engines run on coal. In addition to depleting non-renewable resources, it causes poor air quality, environmental pollution, and severe disruptions to the ecosystem. Land transportation is revolutionized by the introduction of energy or entirely electric engines. Over time, there has been less strain on motor engines.

Throughout history, land transportation has evolved. People everywhere are being exposed to new forms of land transportation. For a better experience, the new line of land transportation adds comfort and luxury while saving time. Transportation encourages travel, which is increasingly becoming a career and hobby.

There are no secret terms related to land transportation; the benefits and drawbacks are covered in detail. Land transportation does, however, have positive effects and make life easier on a daily basis. As land transportation moves toward modernization and innovation, steps have been taken to protect the environment.

Did you get it?

1. **Did land transportation make lives easier?**
2. **What are the popular modes of land transportation?**

5.2. WATER TRANSPORT

Learning Objectives

- Identify different types of water transport
- Discuss market drivers
- Explain market restraints
- Focus on trends and opportunities
- Know water transport market segment analysis
- Describe water transport market regional insights

Commodities, goods, and people have all been transported in significant amounts via waterways. Important commodities that drive the world economy today, such as gas, coal, and oil, are transported by ships. The growth of the water transport market is fueled by the increasing amount of commodities being moved by water. In 2020 alone, approximately 11.2 million metric tons of coal and over 10.9 million metric tons of oil were transported by waterways. It's worth noting that more than 80% of all commodities are shipped by sea, mainly on container ships. Ships are a more cost-effective and suitable option for transporting large, heavy, and bulky goods while generating relatively low pollution compared to other freight transportation methods, thanks to their significant capacity. In 2018, shipping accounted for only 11% of the emissions related to global transportation, much lower than air and land transportation. Water transport is undoubtedly one of the most complex forms of transportation. To ensure a smooth supply chain, various partners are involved, including carriers, ports, tug operators, and freight forwarders.

5.2.1. Types of Water Transport

Airways, waterways, and land routes are the three primary forms of transportation. Air transportation mostly consists of planes and helicopters, whereas land transportation includes everything from cars to buses, trucks to scooters, and bikes to railroads. Other than the different kinds of boats or watercraft, there can be different classifications when it comes to water transportation. Now let's examine a few of them.

Types of Water Transport Based on routes or types of waterways catering to the traffic Based on the waterways, water bodies, and routes catering to maritime or waterborne traffic, water or marine transport can be classified into the following types:

- River

This type of water transportation is specific to a given area and is utilized within rivers. As we all know, rivers are comparatively small freshwater bodies that can have different lengths throughout a nation. Inland waterways also have jurisdiction over riverine transportation. This indicates that only shallow water bodies that take local considerations are allowed to serve the channels or bodies serving the given traffic. Generally speaking, inland water transportation is only intended for quick trips between locations. Similarly, within the borders of a nation or state, river transport refers to the use of small to medium-sized boats for private, public, and commercial purposes for a range of distances.

Keyword

A supply chain includes every step that is involved in getting a finished product or service to the customer.

A medium-sized cargo ship carrying industrial goods or food supplies across vast distances of up to 2000 kilometers can be one of them, or it could be a plain and austere fisherman's boat lounging in the soft dusk sunlight. Among the means of transportation by rivers are barges, fishing boats, speedboats, steamers, trawlers, dinghies, pleasure *yachts*, high-speed planning boats, river ferries, and so forth. All river vessels are categorized as shallow draft vessels due to the restricted or limited depths of the waterbody.

- Lakes

As defined technically, lakes are bounded or enclosed bodies of water with different surface areas. Not only do the areas vary, but so do the depths; they can be as large as the single largest freshwater lake in the world, Lake Superior in the United States, or as small as India's Bhopal Lake. Thus, the quantity and breadth of the lake also influence the kinds of traffic and the means of transportation. Nonetheless, this is also taken into account, practically speaking, when studying inland waterways with low draft.

Keyword

A **yacht** is a sailing or power vessel used for pleasure, cruising, or racing.

- Coastal

Deep-water transportation is responsible for this. Transportation along the coast refers to boats that travel near the coast. They comprise passenger boats as well as commercial boats. As an illustration, suppose there are two ports of call, A and B, separated by 2000 kilometers along a specific coastline. In a short amount of time, say four to five days, a typical coastal vessel makes the journey from point A to point B and back. Commercial coastal vessels include fishing trawlers, coastal coal carriers, and tankers that move coal and petroleum products between two locations in the same nation. Passenger vessels falling under the coastal classification include ferries that charter passengers to travel from a city's port to a well-known beach destination hundreds or thousands of kilometers away (Figure 5.2).

Figure 5.2. Automobile ferry in Croatia.

Source: By Orlovic – Own work, CC BY-SA 3.0, https://commons.wikimedia.org/w/index.php?curid=3032532.

Under classification guidelines, different regulations apply to the design and construction of coastal vessels. They follow their routes as far as they can from the coast and are not required to travel into deep waters. They are claimed to have a substantially smaller range than deep-sea vessels because they are only made to sail a certain distance. Their structural design makes them more durable and stronger than shallow-draft vessels, but far less so than their deep-water equivalents. Shorter distances are also covered by these boats, such as from a coastal area to a neighboring island that might be a well-liked tourist destination.

- International and Intercontinental

International and intercontinental travel accounts for the majority of maritime traffic. Travel across large oceans that separate two continents is known as intercontinental travel. International travel can be between countries that are within the same continental boundary, such as Sweden and Germany or Italy and Greece.

The Atlantic and Pacific, the two largest oceans in the world, see the majority of marine maritime traffic. The strongest and most durable vessels are those made for transcontinental travel. The majority of intercontinental maritime traffic consists of commercial vessels, primarily bulkers, tankers, and containerships, along with defense and research vessels, as a result of aviation's decades-long global monopoly.

- Polar Vessels

They fall within a specific category of vessels. Polar boats are built to operate in frigid waters that contain slabs, floes, bergs, or sheets of ice. They frequently have to overcome the forces and the higher degree of hydrodynamic forces from the ice, so they have extremely strong structural components and are equipped to withstand extremely cold weather for extended periods. Icebreakers are the best kind of polar vessels because they can cut through thicker ice and frequently open a path for a ship that is following.

5.2.1.1. Types of Water Transport Based on Ownership and Utility

Most of the vessels are categorized based on their ownership or utilization per the following categories:

- Commercial

The largest proportion of marine traffic consists of these vessels. Commercial vessels include all types of deep-water or inland vessels used for trade, commerce, and industrial applications. Any type of vessel can be considered a commercial vessel: an LPG carrier, an oil barge, a fishing trawler, or a *containership*. Every commercial vessel is an essential component of the global supply chain and trade at many different levels. Governments or private companies own commercial vessels.

- Passenger and Public Vessels

These boats are designed for public water transportation, primarily for passengers. As mentioned earlier, due to aviation's long-standing dominance, the use of ships for passenger travel has decreased over time. However, they are still popular among vacationers and travelers in various locations for leisure and luxurious recreation. Launches, ferries, steamers, water taxis, and river cruises remain popular choices for short-distance travel worldwide.

- Private Vessels

Keyword

A **container ship** is a cargo ship that carries all of its load in truck-size intermodal containers, in a technique called containerization.

These boats, ranging from small to medium-sized and sometimes larger, are privately owned. They can be owned by an individual or a group, such as an organization. It's important to note that private vessels should not be confused with commercial vessels owned by private individuals. While smaller ships exist, commercial vessels are typically larger, indicating a larger fleet belonging to a shipping company or an international fleet transport company focused on supply chain management and cargo transportation. These vessels operate continuously to transport various goods and supplies over different distances.

For example, a containership owned, managed, and operated by Maersk or an oil tanker owned by Scorpio or Euronav Tankers are examples of mainstream commercial vessels under private or corporate ownership. On the other hand, a medium-sized survey vessel owned and operated by a coastal engineering organization or a few small or medium fishing trawlers operated by a significant seafood processing chain are examples of privately-owned ships. Vessels owned by individuals are primarily for pleasure and recreation, often associated with the affluent class. Yachts are a classic example of such vessels.

- Defense

Among the world's fleet, these ships are also very notable, particularly for superpowers. But every nation has a dedicated naval fleet as part of its military capabilities, regardless of its size, population, or perceived level of global dominance. Defense vessels can be anything from small coast guard patrol boats to fast rescue boats used by the Navy, including warships, carriers, *frigates*, destroyers, and corvettes. A vital component of the naval fleets of many countries are submarines.

- Special Purpose Vessels

These vessels are mainly research or exploration vessels used for scientific purposes. They may be government-owned or operated by institutions or bodies. Last but not least, water transport can be based on submersibility:

- Surface Vessels

All forms of ships and boats that stay afloat in water are surface vessels. They may be further categorized into the following types: displacement, planing, and semi-planing.

- Submarines and submersibles

As we already know, these vessels remain submerged underwater below a certain height. For a more in-depth understanding of submarines and submersibles, read on.

Keyword

A **frigate** is a type of warship. In different eras, the roles and capabilities of ships classified as frigates have varied somewhat.

5.2.2. Market Drivers

The market for water transport has been driven by the need for reasonably priced freight shipping. For heavy items, water freight is usually regarded as inexpensive and reasonable due to the difference in fare pricing when compared to other modes of transportation. A common method of calculating the cost of water transportation is cubic meters of the standard container. The weight and size of the cargo are both taken into consideration when calculating the charged weight for air transportation, in contrast. Products valued at approximately $736 billion were shipped by ocean boats in 2000, according to the U.S. Transport Department. It is projected that foreign trade within the United States will triple by 2022, with the majority of that trade traveling by ocean. One ton of freight can be transported over 100 kilometers via roads for about $3.07, via railways for about $1.93, and via waterways for about $1.62, according to the World Bank, a major financier of India's National Waterway Project. This means that waterways are the most economical method of cargo transportation in India. Water transportation service providers are using sensors more often to keep an eye on ships in far-flung locations.

5.2.3. Market Restraints

Crude oil price volatility is affecting water transport, and it is projected to constrain the water transport market throughout the forecast period. According to the United Nations Conference on Trade and Development, fuel oil is a primary source of energy that powers the worldwide economy, accounting for 95% of the energy utilized in global transportation. Water transport accounts for more than 80% of global goods traffic by volume and is heavily reliant on oil. As the price of oil rises, carriers are compelled to boost their prices or face losses. Typically, the recipient is charged extra to cover the additional expenditures. Product inflation is caused by higher fuel prices, which influence all aspects of product transportation. When fuel costs fall, consumers obtain lower-cost items, while *logistics* businesses use the savings to enhance their operations. Changes in fuel prices have a direct influence on operating expenses. According to US Energy Information Administration statistics, crude oil prices surged in 2022, with Brent crude oil spot prices averaging $71/b for the year compared to $42/b in 2020, raising the operational expenses of water transport and hindering the growth of the water transport market.

Keyword

Logistics is the part of supply chain management that deals with the efficient forward and reverse flow of goods, services, and related information from the point of origin to the point of consumption according to the needs of customers.

5.2.4. Trends and Opportunities

The development of information technology systems for improved vessel management is a significant trend in the water transport industry. With a fleet of 95,402 ships, waterways carried nearly 80% of global trade by volume in 2019, according to the United Nations Conference on Trade and Development (UNCTAD). Information technology is used on ships for customer relationship management, warehouse management, queuing systems, fuel optimization, and monitoring of vessel performance. It is also used to recognize scanned copies and pictures of documents. BunkerEx, an online platform for ship owners that assists in locating appropriate bunker ports; Nautilus Labs, which uses data from sensors, manual reports, and market information to create a unified fleet intelligence; and others are used for ship operations management. As a result, the mix of digital and physical connection assists carriers and seaports in integrating procedures with shippers and tracking devices for containers and cargo.

5.2.5. Water Transport Market Segment Analysis

Inland water freight transport is anticipated to expand at a compound annual growth rate (CAGR) of 5.8% in the water transport market by type, from $16.83 billion in 2022 to $17.80 billion in 2023. Short-term economic recovery from the COVID-19 pandemic was hindered by the Russia-Ukraine war, at least temporarily. A spike in commodity prices, supply chain disruptions that have affected multiple global markets, and economic sanctions against several nations have all been brought about by the conflict between these two nations. Regarding inland water freight transport, the largest region in 2022 was Western Europe. The Asia Pacific region ranked second in terms of *inland water freight transport*. 544.43 million tons of goods were transported over European inland waterways in 2022, an increase of 32.9% from the year before.

Notwithstanding a slight decrease in miles traveled, the total amount of transport completed reached 136 billion ton-kilometers, an increase of 3.3% over the previous year. Inland rivers primarily carried refined petroleum products, coke, and metal ores, among other mining and quarrying materials. Inland waterway transit in Europe is highest in the Netherlands and Germany, encompassing all product categories. Russia has a lot of potential for the development of all forms of transportation due to its geographic location and vast national area. The nation has invested varying amounts in inland transportation infrastructure over the past ten years; in 2020, it is projected to be worth 12.4 billion euros. With almost 11.1 billion metric tons of goods transported in the first quarter of 2022 alone, the road sector carried the most freight.

Keyword

Inland water transport refers to the transportation of people, goods, and materials via waterways such as rivers, canals, lakes, and other navigable bodies of water that are located within a country's boundaries.

Question

What segments are covered in the Water Transport Market report?

Answer

The segments covered in the Water Transport Market report are based on Type, Application, and Region.

5.2.6. Water Transport Market Regional Insights

In 2022, Asia Pacific was the largest region in the water transport market. The region is expected to grow by xx% during the forecast period. Water transport accounts for approximately 95% of India's trade volume and 70% of its trade value. India has 12 major ports and 205 designated small and intermediate ports. Six new mega ports will be built in the country as part of the National Perspective Plan for Sagarmala. Foreign Direct Investment (FDI) of up to 100% is permitted under the automatic route for port and harbor building and maintenance projects. It has also permitted a 10-year tax break for businesses that construct, manage, and operate ports, inland waterways, and inland ports. In FY20, India's principal ports had a capacity of 1,534.91 million tons per year (MTPA). In FY21, all major Indian ports handled 672.60 million tons (MT) of cargo volume. Russia's inland waterway freight volume has experienced significant recent fluctuations, growing by 11% in 2023. Regarding marine transport, coal led in dry cargo by volume and crude oil led in liquid cargo in Russia. Fossil fuels also made up the majority of transshipments in the nation. International exports made up more than one-third of the air freight volume in Russia. In terms of freight tonnage transported in 2022, Air Bridge Cargo Airlines, a member of the Volga-Dnepr Group and a significant national cargo airline, was in first place. The United Kingdom has vast water and wastewater infrastructure, including a network of pipelines spanning hundreds of kilometers and more than 1,000 treatment plants supplying drinking water. More than 25 million houses in England and Wales are connected to public water supply networks and sewerage. However, due to their antiquity, many of these pipes from the Victorian era require maintenance. Every day, three billion liters of water are lost in England and Wales. In addition to infrastructure expenditures, smart water meters will continue to be pushed out across the UK, allowing water companies to monitor water flow and identify leaks much more effectively.

Did you get it?

1. **Which region is expected to hold the highest share in the Water Transport Market?**
2. **What is the market size of the Water Transport Market by 2029?**

5.3. AIR TRANSPORT

Learning Objectives

- Identify characteristics of air transport
- Explain types of air transport
- Describe air freight transport
- Discuss international air transport
- Analyze the importance of air transport
- Know the advantages and disadvantages of air transport
- Understand air transport infrastructure
- Identify an example of ae main airport in the world
- Describe the impact of air transport on the environment

Air transport is the activity that enables the transportation of people, goods, and mail via aircraft. Consequently, the mode of transportation being discussed is airborne and is commonly found in airports.

5.3.1. Characteristics of Air Transport

Airports are the physical infrastructure necessary for organizing this mode of transportation. Therefore, these facilities are essential for ensuring safe aircraft takeoff, landing, and parking. Moreover, they have similar functions to ports for loading and unloading goods, as well as for passenger embarkation and disembarkation. Additionally, they provide the space required for maintenance and refueling.

Air freight transport offers various benefits and characteristics, such as speed, cargo security, dependability, and access to remote locations. Its disadvantages include potential bureaucratic or legal restrictions, increased costs, and technical constraints related to the size and capacity of the aircraft.

5.3.2. Types of Air Transport

Did you know?

The main air powers of the world are the USA, Russia, Japan, Great Britain, France, Canada, and Germany.

Air transportation can be categorized in a variety of ways, starting with the broadest categories that can be used to differentiate between the various types of air transportation (helicopter, plane, airship, hot air balloon, etc.) to more focused categories that group different air transportation modes based on the kind of cargo being transported. In this instance, three sizable groups are identified.

- *Mixed aircraft:* carry passengers in the cabin and goods in the hold. The best-known are those built by Airbus and Boeing.
- *Cargo planes:* transport exclusively goods.
- *Super transports:* aircraft designed to transport large goods.

5.3.2.1. Air Freight Transport

The logistics industry has advanced significantly with the advent of air cargo transport, which is the most recent mode of freight transportation. A mode of transportation that experiences an increase in turnover every year. These days, it has emerged as one of the most widely used methods for moving products for business. In recent years, there has been a noticeable increase in the use of cargo planes for the transportation of goods. A speed that balances the increased expense over other modes of transportation, like a boat. Furthermore, aircraft engines are becoming more and more fuel-efficient due to the rapid advancement of aeronautical technology. That's why using them for international goods transportation is becoming more and more beneficial and profitable.

5.3.2.2. International Air Transport

When compared to other modes of transportation like land or sea, the primary advantage of international air travel is its ability to cover large distances faster. International air travel is also one of the safest ways to travel right now, despite the fact that it costs more than other options. IATA, or the International Air Transport Association, is a global organization that unites the entire aviation sector. The ability of airplanes to travel at high speeds has made international air travel essential for both the global exchange of goods and people.

The ability to travel great distances in a matter of hours enables any exporting or importing business to send or receive cargo, *merchandise*, or goods that could be essential for the supply of a population or an industry, for example. Furthermore, it is feasible to reach markets across the globe thanks to international aviation. Today's logistics chains depend heavily on airplanes because they are the quickest mode of transportation for the growth of international trade operations.

A tight flow and constant supply are hallmarks of most large productive industries, including the automotive industry, in order to prevent or minimize the storage of parts, components, or goods. These industries require goods to be delivered precisely when they need them.

International air transport services facilitate the dispatch and quick shipment of goods by air to any location in the world, as well as continuous just-in-time supply for industries. Almost all industrial sectors and distribution chains use air transport today. Most businesses utilize air transportation to ship samples and paperwork relevant to international trade operations, as well as to market goods and products abroad.

Keyword

Merchandise refers to any type of goods sold to the public. It may include things given away for free by companies aimed at cultivating loyalty.

5.3.3. Importance of Air Transport

The economic benefits of air travel include its capacity to create and sustain jobs, improve international trade and connectivity, encourage travel, and link isolated communities. There is no denying aviation's positive social and economic effects, and all nations—developed and developing—should be interested in seeing the industry flourish. These advantages do, however, come with a price for the environment. For aviation to expand responsibly, the sector must strike a balance between the advantages of increased air travel and its obligation to combat climate change. In addition to facilitating social inclusion and connection, air connectivity also fosters the flow of ideas and knowledge. Likewise, it fosters greater creativity, enhanced productivity, and efficiency.

Remember

All communication routes, transport enterprises and vehicles together form a global transport system.

5.3.4. Advantages and Disadvantages of Air Transport

The primary benefits and features of air freight transportation are its speed, dependability for cargo security, and unrestricted access to remote areas. However, it has disadvantages or drawbacks in terms of increased cost, potential legal or bureaucratic restrictions, and technical limitations because of the aircraft's size and capacity.

Advantages:

- Speed: it is the fastest mode of transport that exists and, therefore, it is especially recommended when time is an important factor.
- There are no physical barriers: thanks to this it is possible to make a trip without interruption choosing the shortest and most direct route through seas, and mountains.
- Easy access: Air transport can transport goods to areas that are not easily accessible by other means of transport.
- Suitable for transporting high-value or perishable goods over long distances.

Disadvantages:

- Very expensive economically: it is the most expensive means of transport.
- Uncertain: air transport is largely conditioned by weather conditions. Snow, rain, fog, etc., can cause the cancellation of scheduled flights and the suspension of air service.
- Not suitable for cheap and high volume goods due to its limited capacity and high cost.
- Legal restrictions: Many countries have legal restrictions in the interest of their own security.

5.3.5. Air Transport Infrastructure

Airports are a vital component of the air transportation system because they offer the support systems required to enable the transfer of people and goods from the ground to various aircraft types as well as the takeoff and landing of aircraft, and touchdowns. Runways, taxiways, terminals for passengers and cargo, aircraft parking lots, boarding bridges, and ground transportation interchanges comprise the fundamental airport infrastructure.

Air travel is currently one of the industries in the global economy that is expanding the fastest. Airports are faced with a challenge in terms of capacity development and management due to this rapid increase. Similarly, the airport industry has changed over the past few decades, since airport infrastructures functioned similarly to other public services. It is now a dynamic and competitive sector, and the businesses that make up this sector have started to leverage their commercial side and, as a result, function as businesses that must make money. That is, to perform tasks more effectively.

5.3.6. Example of a Main Airport

Of the 46 airports and 2 heliports managed by AENA, the main airports in Madrid and Barcelona cover more than 40% of the country's total passengers. At a European level, Spain ranks as the third country in the European Union in the number of air passengers thanks to the airports of Madrid, Barcelona, and Palma de Mallorca. The Adolfo Suárez Madrid-Barajas airport is the gateway between Europe and Latin America (26% seats offered); Madrid is the capital of Spain, and the Adolfo Suárez Madrid-Barajas airport is the gateway for tourism in the country. The Adolfo Suárez Madrid-Barajas airport is connected to most of the country's tourist destinations (28 Spanish airports), with 192 international destinations (+5,000 passengers) from 73 countries, and has 83 airlines (+5,000 passengers), with a presence in all alliances: One World, Sky Team, and Star Alliance. Josep Tarradellas Barcelona-El Prat airport is connected to 212 destinations

(+5,000 passengers), of which 137 are European destinations and has 89 airlines (+5,000 passengers), with all alliances present: One World, Sky Team, and Star Alliance.

5.3.7. Impact of Air Transport on the Environment

The amount of greenhouse gases released into the atmosphere by human activity that are attributed to the commercial aviation sector is estimated to be 2%, and 3% if other polluting gases are taken into account. Over time, this effect will grow. Based on Airbus projections, there will be twice as many passenger aircraft in 2034 as there were in 2014, totaling 35,749 units (of which over 31,000 are still to be built).

Remember

A carbon footprint is the total greenhouse gas (GHG) emissions caused directly and indirectly by an individual, organization, event or product.

5.3.7.1. Solutions to Mitigate this Problem

In order to meet the challenges brought about by its expansion and its effect on the sustainability of the planet, the aviation industry needs to make a strong commitment. Aircraft manufacturers, airlines, airport managers, and ancillary businesses are primarily to blame for this conflict.

In order to reduce its environmental impact, Iberia, to name two specific airline cases, creates various programs aimed at increasing the efficiency of carbon emissions. Improvements to the fleet that include the Airbus A-300–200, which is utilized on long-haul international routes and results in a final kerosene savings of fifteen percent per plane, or improvements to passengers, ground equipment, and aircraft assistance vehicles are examples of projects that help reduce the impact of climate change.

In addition to reducing fuel and CO2 emissions, other initiatives like electronic flight documentation (EFD) and electronic flight bags (EFB) on short- and medium-haul aircraft have improved technology from an operational perspective because the paper documentation portfolios that were in the cabins have vanished, causing the plane's weight to decrease.

On the other hand, the historic milestone achieved by easyJet in reducing CO2 emissions is also worth mentioning. This airline has set itself the ambitious challenge of reducing carbon emissions to 77 grams in 2020, which would mean a reduction of more than 33% in twenty years. For the first time in its history, it has managed to reduce carbon emissions below 80 grams thanks to technological and efficiency improvements in its operations. Since

2000, when the airline started reporting carbon emissions, easyJet has reduced carbon emissions by more than 31% (from 116.2 to 81.05 grams per passenger kilometer last year to 79.98 grams today).

Question

What are the uses of the global air transportation network?

Answer

Air transport provides the only global transportation network for passengers and cargo, making it essential for global business operations and for personal travel and tourism. It provides substantial economic benefits by connecting people and businesses around the globe.

Did you get it?

1. **What is the global air transportation network?**
2. **What are the advantages of air transport?**

5.4. MAJOR ISSUES IN TRANSPORT AND COMMUNICATION

Learning Objectives

- Addressing the negative consequences that unsustainable transportation can have on climate, air pollution, and road safety
- Understand climate-proofing transport infrastructure
- Know that the world's logistics systems are struggling to keep up with the times
- Understand why the world needs new ideas and new ways of integrating transport modes
- Financing sustainable transport

Access to markets, jobs, healthcare, education, and education are all made possible by transportation, which is a critical enabler of economic growth. Apart from the aforementioned advantages for development, enhancing transportation networks through increased cleanliness, safety, inclusivity, and resilience spurs local and global climate action. Nevertheless, despite the importance of transportation to key development goals, there are still large gaps that impede development and the benefits that improved transportation can bring to reducing *poverty*, promoting economic growth, and achieving climate goals.

Governments in developing nations, local governments, multilateral development banks, and the commercial sector should work together to promote more environmentally friendly transportation options in these regions. To achieve this, developing nations need to address the following five major transportation-related issues.

Addressing the negative consequences that unsustainable transportation can have on climate, air pollution, and road safety is crucial. Approximately 25% of global CO2 emissions come from the transportation sector. Moreover, over 7 million people die each year due to complications from air pollution, and an additional 1 million people die in car accidents. Countries must find ways to reduce transportation-related emissions while also improving access to economic opportunities to advance broad-based sustainable development.

This framework can be implemented in various ways by nations to lower emissions and enhance mobility. Perhaps one of the most effective methods to increase mobility and reduce pollution and carbon emissions simultaneously is by improving public transportation. Enhancing the quality and coverage of public transportation systems and integrating them into initiatives to promote safe cycling and walking in well-planned cities can bring additional benefits. Electric mobility, including technologies like electric bikes, rickshaws, and buses, is now affordable for many developing nations. Governments can encourage the use of these vehicles by building charging infrastructure and devising innovative financing solutions to keep costs low.

Keyword

Poverty is a state or condition in which a person or community lacks the financial resources and essentials for a minimum standard of living.

- **Climate-proofing transport infrastructure.**

Transport systems play a crucial role in the functioning of our world, but they are exposed to significant risks from both natural and man-made disasters. In today's unstable world, characterized by unpredictable climate-related catastrophes, pandemics, and human-induced crises, transportation often becomes a victim. With the escalating dangers posed by climate change, countries must be prepared for the mounting expenses associated with transport infrastructure. Research conducted by the World Bank provides an estimate that the cost of maintaining roads in Africa will surge by 270% in the forthcoming decades due to the impacts of climate change.

Regretfully, for many people, particularly those in developing nations, experiencing crises seems to be the new normal. Transportation investments are no exception to the need for greater resilience on the part of the global community. When a calamity occurs, the physical infrastructure must endure. Governments can save a lot of money by designing with the future in mind and avoiding the need for a rebuild or retrofit. This holds true for many things, like enhanced emergency readiness, building redundancies in logistical networks, and making sure that assets are managed well, like maintaining roads.

- The world's logistics systems are struggling to keep up with the times.

The world's logistics systems are incredibly fragile, as COVID-19 shockingly demonstrated. Global shortages of food and other necessities were caused by unstable energy prices, traffic jams at major ports, and geopolitical unrest.

By volume, more than four-fifths of all merchandise trade worldwide is transported by sea, and container trade has been expanding more quickly than the world GDP overall. Any meaningful change in global logistics must consider the maritime industry. To streamline operations, allow for real-time tracking, and make data sharing between shipping lines, port services, cargo handling operations, and clearing agencies with other transport networks easier, digitizing port processes is essential. Policies need to be changed by governments in some areas. The private capital required to finance infrastructure projects at ports and elsewhere would be generated with the aid of an improved investment climate. Additionally, in order to improve market competitiveness and cut costs, state-owned enterprises must undergo reform. Logistics firms should also begin preparing for the switch to more environmentally friendly fuels, like hydrogen, to power ships and aircraft, in order to protect themselves against future fluctuations in fuel prices.

Remember

High initial investment, technological obsoletion, integration complexity, and scalability are common challenges in implementing communication systems for transportation.

- The world needs new ideas and new ways of integrating transport modes.

Cities are growing – 2.5 billion people will be added to urban areas by 2050, most of them in low- and middle-income countries. Governments need to ensure their transport systems keep up.

Technology breakthroughs are providing answers like smarter batteries, open data on public transportation, real-time air quality sensors, and the ability to link various forms of transportation using AI and data. This kind of integration has the power to drastically alter how people travel. All of these tools indicate a bright future for transportation, but only if it is available to all. To promote a future of data-informed policymaking worldwide, decision-makers must have access to the newest tools. Enhancing technical capabilities and governance will facilitate the efficient utilization of technology by the rapidly expanding small and medium-sized cities, allowing them to establish integrated transportation networks and facilitate smooth connectivity among various modes for all users.

- Financing sustainable transport.

Governments have historically provided funding for extensive infrastructure systems, but public funds are scarce. Over half of developing nations are on the brink of financial distress, and public debt is at an all-time high.

It will take the cooperation of numerous parties outside of government, particularly the private sector, to reach the investment levels required to establish contemporary, safe, and efficient transportation systems. The money is there: in 2021, the 500 biggest asset managers globally managed a record $119 trillion in assets. However, the uncertainty surrounding the world has increased risk, which discourages private investment in infrastructure projects.

This is where the World Bank and other development banks come into play, helping to kickstart projects with creative financing arrangements. Research, analytics, and convening can be provided in tandem by WRI and other stakeholders. Development banks can work with national governments and cities to encourage more private investors to support the necessary changes to vehicle fleets and transport infrastructure by combining concessional finance, public-private partnerships, and other risk-mitigating tools, like guarantees.

Question

What is ecotourism?

Answer

Ecotourism typically involves travel to destinations where plants and animals thrive in a naturally preserved environment. The Amazon rainforest, African forest safari, and trekking in the slopes of the Himalayas are famous incredible eco-friendly attractions.

Did you get it?

1. **What are the problems faced by the transport industry?**
2. **What is the carbon footprint of transportation?**

ROLE MODEL

GEORGE CAYLEY: THE FATHER OF AERONAUTICS

Sir George Cayley (born December 27, 1773, Scarborough, Yorkshire, England—died December 8, 1854, Brompton, Yorkshire) was an English pioneer of aerial navigation and aeronautical engineering. He was the designer of the first successful glider to carry a human being aloft.

Fascinated by flight since childhood, Cayley conducted a variety of tests and experiments intended to explore aerodynamic principles and to gather information of value in the design of aircraft. He published the results of his original research in 1809. His most important discoveries included the advantages of streamlining, the means of obtaining longitudinal and lateral stability, elements of wing design, thoughts on biplane and multiplane wings, and the use of rudders and elevators for control. Throughout his active career, Cayley designed a variety of aircraft, including helicopters, airships, and fixed-wing machines.

Cayley established the modern configuration of an airplane as a fixed-wing flying machine with separate systems for lift, propulsion, and control as early as 1799. In 1804, he flew the first successful glider model for which there is no record. His work culminated in 1853 with the completion of a full-scale glider that carried his reluctant coachman on the first manned glider flight on record.

An individual of wide technical and scientific interests, Cayley invented the light-tension wheel (forerunner of the bicycle wheel), the expansion-air, or hot-air, engine (1805), and the caterpillar tractor (1825). He was a founder of the Regent Street Polytechnic Institution (charter of incorporation granted in 1839; now the Royal Polytechnic Institution). He also pursued research in science education, land reclamation, acoustics, railway equipment, lifeboats, ballistics, optics, and electricity.

FOCUS ON CAREERS

TRANSPORTATION ANALYST

A transportation analyst specializes in analyzing transportation systems and identifying ways to improve their efficiency. These analysts work with government agencies, transportation companies, and logistics firms to develop strategies that optimize transportation while minimizing costs. These professionals are responsible for analyzing traffic patterns, assessing the impact of new transportation initiatives, and recommending improvements for efficiency and sustainability.

What Does a Transportation Analyst Do?

A transportation analyst specializes in performing research and analysis to optimize transportation processes, ensuring convenience and motorist safety. They are primarily in charge of identifying the strengths and weaknesses of current systems, recommending and implementing solutions to problem areas, and performing risk assessments. A transportation analyst may also participate in distributing budgets, maintaining databases, training new workforce members, developing new programs and projects, implementing the company's policies and regulations, and creating new ones when necessary.

Transportation Analyst Pros and Cons

Pros:

- Opportunity to work in a growing and dynamic industry
- Possibility to work with cutting-edge technology and data analysis tools
- Potential for high pay and job security
- Varied work tasks and projects, providing intellectual stimulation
- Possibility to work remotely or from multiple locations

Cons:

- Work can be demanding and require long hours
- High pressure to meet deadlines and produce quality work
- Can require extensive travel or fieldwork, which may be physically demanding
- May require significant technical expertise or experience, limiting entry-level job opportunities
- Potential for limited career mobility within the field

A CLOSER LOOK

ECOTOURISM

Ecotourism involves traveling sustainably. When you vacation, domestically or abroad, you stay conscious of the environment as much as possible. Ecotourists try to limit their carbon footprint and support local ecosystems by contributing positively. For example, they could eat at a local restaurant or refrain from using plastic on their trip. Ecotourism has become critical as people try to expand their worldview while staying environmentally conscious.

Tourism is vital for many communities worldwide. Vacationers spend their money to help small businesses thrive and to stimulate local economies. However, tourism can negatively impact the environment. A 2022 study found that tourism is responsible for nearly 8% of the world's carbon dioxide (CO_2) emissions, most of which came from food and waste management. The pandemic hit the tourism industry with a steep 74% decline in international travel. Many countries had lockdowns in place to prevent COVID-19 transmission from foreign places. While this contributed to a significant drop in carbon emissions in 2020, tourist-dependent nations suffered huge economic losses.

Three years after the first case was detected in Wuhan, China, the pandemic is finally winding down and international travel is resuming, with air traffic set to reach 2019 levels soon. Last year saw a 153% increase in air travel compared to 2021 and about 62% pre-pandemic levels. In 2023, air traffic is expected to continue rising as most countries lift restrictions. The bounce-back of tourism means the same will happen in terms of emissions. In 2022, GHG emissions increased by 7% in the first quarter compared to 2020.

It is critical to practice ecotourism as global warming becomes more apparent. You will benefit from learning and becoming a better friend to the environment. Implementing ecotourism comes with many benefits:

- Educate yourself: The most significant benefit of ecotourism is educating yourself on environmental issues. If you find ways to be environmentally conscious on your trip, you are more likely to repeat those behaviors at home. Reading literature and research from scientists puts into perspective how the planet needs your help to survive.
- Protect resources: Sustainable travel means using natural and renewable resources to improve the planet's health. You are protecting the environment around you from the negative impacts of travel. If you believe in leaving something better than when you found it, ecotourism is the way to go.
- Help economies: Practicing ecotourism means other vacationers behind you also get to enjoy a clean environment. One way is by spending your money at local businesses. These shops are the heart of communities and give the location its remarkable personality. Patronizing these companies helps them grow and continue their services for future visitors.

SUMMARY

- The movement of people, animals, goods, and services is facilitated by land transportation. Land transportation is available in various forms and has transformed the concept of movement on a global scale. In the past, walking served as the primary mode of transportation. The range of transportation options has expanded over time.
- To transport passengers, manufactured goods, food, and raw materials, strong animals have been utilized. Animals have reduced human labor and effort by covering the same distance in half the time. The invention of wheels and carts has enhanced accessibility for individuals seeking to transport goods.
- The two primary categories of land transportation are railroads and roads, with road transportation being the oldest mode.
- The impacts of land transportation are visible in people's daily lives worldwide.
- Commodities, goods, and people have been transported in large quantities through waterways.
- Airways, waterways, and land routes are the three main forms of transportation. Air transportation mainly involves planes and helicopters, while land transportation encompasses a variety of vehicles such as cars, buses, trucks, scooters, bikes, and railroads.
- The demand for affordable freight shipping has been a key driver in the water transport market. For heavy items, water freight is usually regarded as inexpensive and reasonable due to the difference in fare pricing when compared to other modes of transportation.
- Crude oil price volatility is affecting water transport, and it is projected to constrain the water transport market throughout the forecast period.
- The development of information technology systems for improved vessel management is a significant trend in the water transport industry.
- Inland water freight transport is anticipated to expand at a compound annual growth rate (CAGR) of 5.8% in the water transport market by type, from $16.83 billion in 2022 to $17.80 billion in 2023.
- Asia Pacific was the largest region in the water transport market in 2022. It is projected to grow by xx% during the forecast period.
- Air transport enables the movement of people, goods, and mail by aircraft. This mode of transportation is airborne and operates in airports. Airports provide the necessary infrastructure for organizing this transportation mode.
- Air transportation can be classified in different ways, starting with broad categories that help distinguish between various types of air transportation like helicopters, planes, airships, hot air balloons, etc.) to more focused categories that group different air transportation modes based on the kind of cargo being transported.

- The economic benefits of air travel include its capacity to create and sustain jobs, improve international trade and connectivity, encourage travel, and link isolated communities.
- The primary benefits and features of air freight transportation are its speed, dependability for cargo security, and unrestricted access to remote areas.
- Transportation enables access to markets, jobs, healthcare, and education, crucial for economic growth. Improving transportation networks with cleanliness, safety, inclusivity, and resilience drives local and global climate action, benefiting development.

MULTIPLE CHOICE QUESTIONS

1 Which mode of transport is most commonly used for the movement of goods within cities?

a. Air transport
b. Water transport
c. Road transport
d. Rail transport

2. What is the term for the transfer of passengers and goods from one mode of transport to another during a journey?

a. Transfer station
b. Multimodal transport
c. Transport junction
d. Transit hub

3. What type of transport uses underground tunnels to move people and goods?

a. Water transport
b. Rail transport
c. Road transport
d. Subway transport

4. Example of road transportation are

a. walking
b. cycling
c. animal carts
d. driving a car
e. all of the above

5. Heavy traffic can increase travel time and may cause a delay in arrival at the destination.

a. True
b. False

6. **A medium-sized cargo ship carrying industrial goods or food supplies across vast distances of up to _______ can be one of them, or it could be a plain and austere fisherman's boat lounging in the soft dusk sunlight.**
   a. 1000 kilometers
   b. 500 kilometers
   c. 2000 kilometers
   d. None of the above
7. **________ are bounded or enclosed bodies of water with different surface areas.**
   a. Lakes
   b. Coastal
   c. International and Intercontinental
   d. None of the above
8. **Polar boats are built to operate in frigid waters that contain slabs, floes, bergs, or sheets of ice.**
   a. True
   b. False

REVIEW QUESTIONS

1. How many types of transportation exist in today's world?
2. What is the role and importance of transportation and communication?
3. What are the means of transportation and communication called?
4. What is the forecast period for the Water Transport Market?
5. What was the market size of the Water Transport Market in 2021?
6. Who invented the first air transport?
7. How much carbon emissions come from the transportation sector?

Answer to Multiple Choice Questions

1. (c) 2. (b) 3. (d) 4. (e) 5. (a)

6. (c) 7. (a) 8. (a)

REFERENCES

1. Babar, Y., & Burtch, G., (2020). Examining the heterogeneous impact of ride-hailing services on public transit use. *Information Systems Research*, *31*(3), 820–834. https://doi.org/10.1287/isre.2019.0917.
2. Briassoulis, H., (2020). *Analysis of Land Use Change: Theoretical and Modeling Analysis of Land Use Change: Theoretical and Modeling Approaches.* WVU

Research Repository. https://researchrepository.wvu.edu/cgi/viewcontent.cgi?article=1000&context=rri-web-book (accessed on 04 April 2024).

3. Carter, E. C., Chadda, H. S., & Schonfeld, P. M., (1984). Comparison of transportation planning in developed and developing countries. *Transportation Quarterly*, *38*(1), 69–86. https://trid.trb.org/view/204191 (accessed on 04 April 2024).
4. Cervero, R., (2006). Alternative approaches to modeling the travel-demand impacts of smart growth. *Journal of the American Planning Association*, *72*(3), 285–295. https://doi.org/10.1080/01944360608976751.
5. Cervero, R., Rood, T., & Appleyard, B., (1999). Tracking accessibility: Employment and housing opportunities in the San Francisco Bay Area. *Environment and Planning A*, *31*(7), 1259–1278. https://doi.org/10.1068/a311259.
6. Connors, W., (2002). *By the Next Boat: A Photo History of Newfoundland Coastal Boats*. St. John's: Johnson Family Foundation.
7. Cuff, R. H., (2003). *Locomotives Since 1880*. St. John's: Railway Coast Museum.
8. Earle, G., (2003). *How We Got Around*. St. John's: Railway Coast Museum.
9. Penney, A. R., & Fabian, K., (2003). *A History of the Newfoundland Railway*. St. John's: Harry Cuff Publications Limited.

CHAPTER 6

Demography and Census

LEARNING OBJECTIVES

After studying this chapter, you will be able to:

- Understand the concept of human resources
- Discuss population growth
- Explain demographic dividend
- Identify components of population change

"Individual cultures and ideologies have their appropriate uses but none of them erase or replace the universal experiences, like love and weeping and laughter, common to all human beings."

—*Aberjhani*

INTRODUCTORY EXAMPLE

Age Structure Dynamics

The primary indicators of a population's age structure are its history of fertility and mortality. Improved survival usually happens disproportionately in children in high-mortality populations. As a result, there is a baby boom. When desired fertility decreases with economic development and fertility abates in response to perceptions of improved child survival, the fertility boom eventually comes to an end. However, as the comparatively large baby-boom cohorts enter adulthood and pass through adolescence, the percentage of the population at the prime ages for saving and working increases.

As a result, the economy's potential for rapid income growth and the eradication of poverty is increased on a per capita basis. Recent events demonstrate that nations that are unable to produce enough jobs for sizable populations of young adults are vulnerable to social, political, and economic instability. Examples of these events include the Arab uprisings and the more recent large-scale protests in Sudan and Chile.

The "demographic dividend" refers to the process through which a changing age structure can spur economic growth. It depends, of course, on several complex factors, including the nature and pace of demographic change, the operation of labor and capital markets, macroeconomic management and trade policies, governance, and human capital accumulation. Nonetheless, the demographic dividend model can account for significant variation in past economic performance among different countries and regions (e.g., East Asia vs. Latin America vs. sub-Saharan Africa) and helps identify more- and less-promising country settings for future economic growth. For example, from 2020 to 2030, Nepal, Jordan, Bhutan, and Eswatini are projected to experience the largest gains among countries in the ratios of their working-age to non-working-age populations.

The dependency ratio, which is the opposite of the working-age to non-working-age ratio, quantifies the financial burden that people in the working age bracket bear when providing for those who are not in the working age bracket in addition to themselves. The ratio in 1990 showed a significant difference between the more developed and less developed regions (0.68 versus 1.04).

In less developed regions, the ratio dropped to 0.75 by 2020, while in more developed regions it increased to 0.70 due to distinct patterns of declining fertility and aging populations. Furthermore, it is predicted that by 2050, the dependency ratio will be higher in more developed regions (0.89) than in less developed ones (0.77). The aforementioned shift implies that in the upcoming decades, the demographics of less developed regions will be more conducive to economic prosperity than those of more developed regions. Africa is the only continent where a decline in this ratio is predicted by 2050, so this will be particularly true there.

Policies in nations like Chad, the Central African Republic, Somalia, and Sierra Leone that have not yet undergone significant demographic shifts are suitably focused on accelerating those shifts. These policies include funding initiatives that support the survival of infants and children, like increasing vaccination coverage and facilitating greater access to primary healthcare facilities that are adequately staffed and equipped.

Policy measures to facilitate a decrease in fertility, like boosting girls' education and providing access to family planning and reproductive health care, could be advantageous for populations that have seen improvements in health and survival.

In addition, nations with sizeable populations concentrated in the high-work and high-savings stages of the life cycle require policies to fully capitalize on the potential advantages of advantageous demographics. These policies include fostering the health of capital and labor markets, building infrastructure, providing workers with human capital, implementing prudent macroeconomic management, crafting well-thought-out trade policies, and upholding good governance. Though a high proportion of the population is working age, such policies are always desirable.

Although per capita income in some of today's advanced economies is lower in real terms than it was when they were at a comparable demographic stage, investing in these different sets of policies may prove difficult in these countries.

INTRODUCTION

"Demography is destiny" is an oft-cited phrase that suggests the size, growth, and structure of a nation's population determine its long-term social, economic, and political fabric. The phrase highlights the role of demographics in shaping many complex challenges and opportunities societies face, including several pertinent to economic growth and development.

To say that demography determines everything, however, is oversimplified because it ignores the reality that changes in technology, cultural norms, and behavior, as well as institutional and policy reforms, can all have an impact on demographic trajectories and their implications for development.

Global population growth, shifting fertility and mortality patterns, and corresponding shifts in the age composition of the population are the three main drivers of this massive demographic upheaval.

6.1. CONCEPT OF HUMAN RESOURCES

Learning Objectives

- Identify human resources responsibilities
- Explain HR management strategies
- Understand internal vs. external human resources

The workforce or employees of a company are collectively referred to as human resources (HR). Nonetheless, it's also frequently employed to denote the division or team in charge of overseeing an organization's workforce. Human resources management, sometimes referred to as HR or personnel department, is the purview of a dedicated human resources department in many organizations.

In order to effectively recruit, train, manage, and support an organization's human capital, HR professionals are essential. When human resources departments are run well, they support a company's business goals in addition to increasing employee satisfaction.

6.1.1. Human Resources Responsibilities

Depending on the size, industry, and organizational structure of the business, the precise duties assigned to HR can differ greatly amongst organizations. HR departments are usually responsible for maintaining a healthy work environment, supporting employee welfare, enforcing workplace regulations, and assisting the organization in accomplishing its goals through efficient workforce management. HR duties could involve the following:

6.1.1.1. Recruitment and Staffing

Hiring competent applicants to fill open positions within the company is the responsibility of HR. This may involve writing job descriptions, distributing job listings, conducting background checks, scheduling interviews, and overseeing the hiring process. Additionally, HR specialists often negotiate salaries with potential hires and provide compensation recommendations, establishing salary ranges for positions based on experience and location.

6.1.1.2. Employee Onboarding

HR is responsible for facilitating new hire orientation and integration into the company. This includes sharing information about the company's policies, benefits, and culture, as well as standard procedures such as setting up accounts, email addresses, equipment, and business cards for employees. The onboarding process may also include submitting paperwork to government agencies like the IRS and the Bureau of Labor Statistics and maintaining employee records.

Keyword

Employee welfare is a term that encompasses a broad range of benefits and services that an employer may offer to its employees.

6.1.1.3. Training and Development

HR divisions are in charge of training and development initiatives that assist staff members in gaining the knowledge and abilities necessary to advance in their careers. HR specialists may plan on-site training courses in certain situations; in other situations, businesses provide online training through webinars and training apps. In addition, HR might offer outside resources for learning and growth, like admission to a scientific conference or a course.

6.1.1.4. Performance Management

HR departments evaluate employee performance in addition to people managers. Annual or biannual performance cycles may be established, allowing managers and staff to discuss performance, raises, and improvement plans while exchanging formalized feedback. In order to work more closely with employees on their career paths within the organization, certain HR departments may even provide the feedback themselves.

6.1.1.5. Worker Compensation and Benefits Administration

Salary structures, bonuses, health insurance, stipends, and retirement plans are just a few of the benefits and compensation packages that HR departments oversee. Payroll processing is usually the domain of other departments, such as accounting, but HR managers may oversee the mechanism that decides employee raises.

6.1.1.6. Employee Relations

HR can be contacted for assistance with workplace disputes, complaints, and employee issues. It settles conflicts between coworkers and offers assistance to specific employees. In addition to enforcing organizational behavior standards, an HR department strives to promote a positive work environment. Furthermore,

HR departments frequently focus on raising worker morale and *productivity*. HR teams are often responsible for tasks like organizing off-site gatherings, organizing team-building activities, hosting office parties, and giving out holiday merchandise.

Keyword

Productivity is commonly defined as a ratio between the output volume and the volume of inputs.

6.1.1.7. Policy Enforcement and Compliance

The development and enforcement of company policies and procedures that uphold morally and legally sound working conditions are aided by human resource management teams. They guarantee adherence to regional, national, and local regulations concerning equal employment, harassment, discrimination, and workplace security. HR departments are required to maintain a current awareness of laws that affect their workers and promptly create policies that comply with these laws.

6.1.1.8. Employee Records and Data Management

HR keeps track of employee data, including personal information, performance reviews, and payroll records, to support different HR procedures. When it is lawful and suitable, HR personnel may provide employee data to other departments or even to outside companies, such as a 401(k) manager.

6.1.1.9. Organizational Development

In order to match the workforce with the company's objectives and growth strategies, HR may be involved in organizational development and strategic planning processes. HR may occasionally be involved in changes to an organization's organizational structure, such as the creation of new departments or the simplification of managerial positions.

Remember

Payroll and benefits administration are increasingly being outsourced to third-party providers rather than being handled in-house by many companies.

6.1.1.10. HR Technology Implementation

Software programs can take care of a lot of the HR requirements of today. Therefore, in order to handle HR-related data and automate administrative tasks, HR departments frequently make use of technology and software. Using HR technology frees up HR managers to concentrate on more strategic tasks like performance management and organizational development.

6.1.2. HR Management Strategies

Strategic initiatives were implemented within HR departments starting in the 1980s. The foundation of this movement was research on the effects of employee-related issues on a company's long-term financial performance.

Did you know?

In 1901, National Cash Register Co., known today as NCR Corp., established what is thought to have been the first personnel department (the HR predecessor) to manage wages, workplace safety, and employee concerns.

These strategies combined are sometimes referred to as human resource management (HRM) strategies. HRM is a comprehensive management approach for staff members as well as the environment and culture of an organization. It focuses on recruitment, management, and providing overall guidance for employees within an organization.

When an HR department utilizes HRM practices, it typically adopts a more proactive approach to improving the workforce within the company. It may offer advice to management on procedures, strategies, and business solutions.

Google is one of the companies that has taken a proactive stance on employee relations through its HR division. The company offers a wide range of benefits to its employees, including fitness centers, massage therapy, on-site wellness centers, on-site meals and snacks, and ergonomic support. Google considers happy employees to be productive employees.

Question

What is human resource management?

Answer

Human resource management is the strategic and operational oversight of an organization's workforce to ensure it effectively contributes to the company's goals and objectives.

6.1.3. Internal vs. External Human Resources

Companies have the option of establishing an HR department in-house or hiring a third-party company to handle their HR needs. While the latter is known as external human resources management, the former is known as internal human resources management. In a hybrid system, businesses can also outsource certain functions to a third party, including payroll, benefits, hiring, compliance, and training. Positions that necessitate being acquainted with the team, like performance evaluation and recommendations for promotions, handling conflicts, and strategic planning, remain internal.

Did you get it?

1. **What does the human resources department do?**
2. **What are the human resources functions?**

6.2. POPULATION GROWTH

Learning Objectives

- Focus on demographic theories
- Discuss calculating population growth
- Understand population dynamics
- Explain human population growth

Population growth refers to the variation in the number of individuals belonging to a specific plant or animal species in a given area over a specific duration in population ecology. Fertility, mortality, and, in the case of animals, migration are factors that influence population growth. Emigration or immigration to a specific place. The population growth rate is the term used to describe the average change in a population over time. An increase in population is indicated by a positive growth rate, while a decrease in population is indicated by a negative growth rate. The maximum population that can exist in a particular environment is known as the carrying capacity of the environment, and it is based on the quantity and accessibility of resources necessary for the population to survive.

6.2.1. Demographic Theories

Sociologists view population issues as crucial for understanding human interactions. The four main population theories are Malthusian, zero population growth, *cornucopian*, and demographic transition theories.

6.2.1.1. Malthusian Theory

Thomas Malthus (1766–1834) was an English clergyman who made dire predictions about the Earth's ability to sustain its growing population. According to Malthusian theory, three factors would control the human population that exceeded the Earth's carrying capacity, or how many people can live in a given area considering the amount of available resources. Malthus identified these factors as war, famine, and disease. He termed them "positive checks" because they increase mortality rates, thus keeping the population in check. They are countered by "preventive checks," which also control the population by reducing fertility rates; preventive checks include birth control and celibacy. Thinking practically, Malthus saw that people could produce only so much food in a given year, yet the population was increasing at an exponential rate. Eventually, he thought people would run out of food and begin to starve. They would go to war over increasingly scarce resources and reduce the population to a manageable level, and then the cycle would begin anew.

Keyword

Cornucopian is a label given to individuals who assert that the environmental problems faced by society either do not exist or can be solved by technology or the free market.

6.2.1.2. Zero Population Growth

Paul Ehrlich, a neo-Malthusian researcher, carried Malthus's predictions into the twentieth century. Ehrlich contends that the environment—rather than the food supply in particular—will be vital to the planet's population's long-term health. According to Ehrlich's theories, as a result of the affluent population's depletion of or pollution of various natural resources, including air and water, the planet is rapidly approaching total environmental collapse. He promoted the idea of zero population growth (ZPG), which states that the number of people who join a population by birth or immigration equals the number of people who leave it by emigration or death. Even though opinions on this idea are divided, it is nevertheless thought to be a potential remedy for world *overpopulation*.

Keyword

Overpopulation is the state whereby the human population rises to an extent exceeding the carrying capacity of the ecological setting.

6.2.1.3. Cornucopian Theory

Undoubtedly, there are theories that place less emphasis on the gloomy prediction that the world's population will face an unfavorable challenge to survival. Cornucopian theory laughs at the notion that humans will eventually go extinct and claims that human creativity will find a way to solve any problems that may arise in the social or environmental sphere. For instance, it brings up the problem of food supply. The theory argues that agricultural scientists, who have been doing so for centuries, will figure out how to grow more food if we need it. This viewpoint holds that human ingenuity has proven capable of solving problems for thousands of years, and there is no reason why this pattern should stop.

6.2.1.4. Demographic Transition Theory

We can discern distinct trends in population growth, regardless of one's belief that humankind will always adapt to changing conditions or that an environmental catastrophe will wipe out humanity as we know it. The transition from an unindustrialized to a postindustrial society follows a predictable continuum. Future population growth is predicted by demographic transition theory to follow a predictable four-stage model.

In Stage 1, birth, death, and infant mortality rates are all high, while life expectancy is short. An example of this stage is the 1800s in the United States. As countries begin to industrialize, they enter Stage 2, where birth rates are higher while infant mortality and the death rates drop. Life expectancy also increases. Afghanistan is currently in this stage. Stage 3 occurs once a society is thoroughly industrialized; birthrates decline, while life expectancy continues to increase. Death rates continue to decrease. Mexico's population is at this stage. In the final phase, Stage 4, we see the postindustrial era of a society. Birth and death rates are low, people are healthier and live longer, and society enters a phase of population stability. The overall population may even decline. For example, Sweden is considered to be in Stage 4.

Keyword

Fecundity is a measure of the reproductive capacity of an individual or population, typically restricted to the reproductive individuals.

Countries are divided into three categories by the United Nations Population Fund (2008): high, intermediate, and low fertility. High-fertility nations, which are primarily found in sub-Saharan Africa, are expected to see a threefold increase in population between 2011 and 2100, according to UN estimates. The growth rate for nations with intermediate fertility rates—which includes the US, India, and Mexico—is predicted to be roughly 26%. Furthermore, there will be a 20% population decline in low-fertility nations like China, Australia, and most of Europe. The graphs that follow show this pattern.

6.2.2. Calculating Population Growth

Population loss, or the combined rates of mortality and emigration, can be subtracted from population gain, or the combined rates of fertility and immigration, to determine population growth rates in a given area and time period. Fertility is the average number of children that a member of a species produces over time in a given set of environmental conditions. *Fecundity*—the theoretical maximum number of offspring that a species member can produce in a given amount of time—should not be confused with fertility. Because they limit the number of people who can reproduce, a population's age distribution and sex ratio have an impact on fertility rates (Figure 6.1).

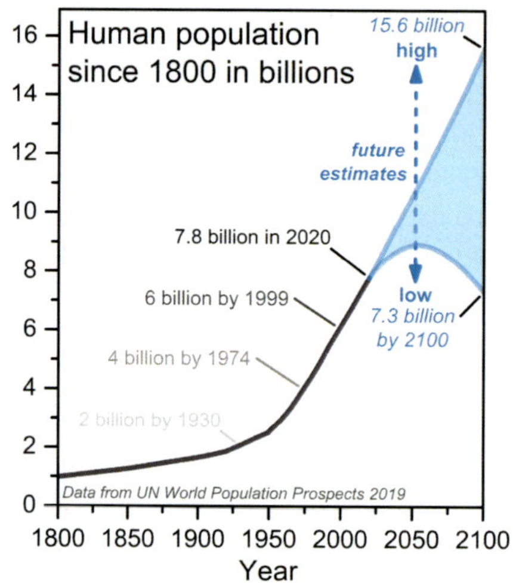

Figure 6.1. World human population estimates from 1800 to 2100, with estimated range of future population after 2020 based on "high" and "low" scenarios.

Source: By Bdm25, Own work, CC-BY-SA-4.0, https://upload.wikimedia.org/wikipedia/commons/thumb/c/c8/Human_population_since_1800.png/786px-Human_population_since_1800.png?20200417195327.

6.2.3. Population Dynamics

Because of its carrying capacity, the natural environment is vital to population growth. The limited life-sustaining resources in an environment can be diminished for individual members of a species due to a variety of factors, including increased population density and competition from other species. Furthermore, the availability of food, mates, and environmental stressors like pollution and natural disasters all have an impact on fertility, mortality, and migration rates.

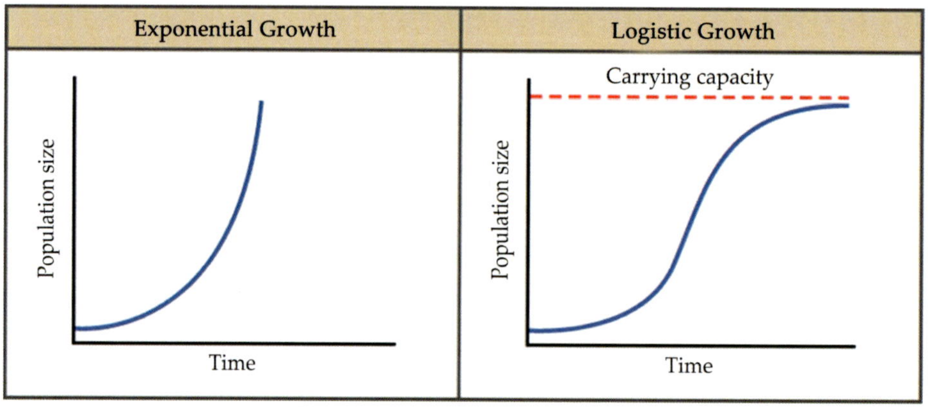

Figure 6.2. Exponential versus logistic population growth.

Source: https://bio.libretexts.org/Bookshelves/Introductory_and_General_Biology/Book%3A_General_Biology_%28Boundless%29/45%3A_Population_and_Community_Ecology/45.02%3A_Environmental_Limits_to_Population_Growth/45.2B%3A_Logistic_Population_Growth.

An S-shaped curve called a logistic curve, as shown in Figure 6.2, can be used to graphically represent the dynamics of population growth. An initial growth lag, an exponential growth spurt, and ultimately a decline in population growth are depicted by the logistic curve. There is a growth plateau at the end of the curve when population density is high because mortality tends to rise due to increased disease transmission, predation, or competition for resources. Seasonal changes in the environment may also have an impact on population growth rates. When a spring thaws in deep lakes, for instance, the colder, deeper waters rise to the surface of the lake and release nutrients, which in turn trigger bursts in the growth of plankton, including bacteria, protozoa, and algae.

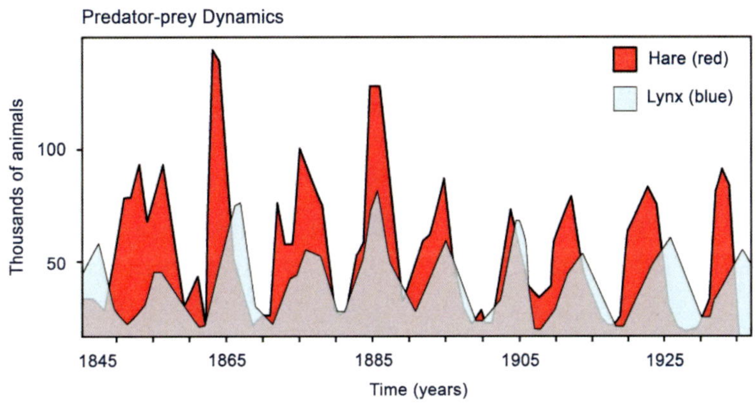

Figure 6.3. Hare and lynx populations over time. Cyclical fluctuations in the population density of the snowshoe hare and its effect on the population of its predator, the lynx.

Source: https://courses.lumenlearning.com/suny-monroe-environmentalbiology/chapter/4–4-community-ecology/ Community Ecology by OpenStax is licensed under CC BY 4.0. Modified from the original by Matthew R. Fisher.

Predator-prey relationships can also affect how quickly a species' population grows. For instance, when the number of zooplankton, or microscopic aquatic animals, rises, the population of their algae prey may either decline or grow more slowly. This leads to an oscillating pattern of population, like the cyclical fluctuations of the lynx and snowshoe hare populations, as shown in Figure 6.3.

Question

What is logistic population growth?

Answer

The logistic model takes the shape of a sigmoid curve and describes the growth of a population as exponential, followed by a decrease in growth, and bound by a carrying capacity due to environmental pressures.

6.2.4. Human Population Growth

The number of people on the planet has been growing exponentially since the 18th century, but growth rates have been decreasing since the second half of the 20th century. The overall rate of growth is declining, although population growth rates vary significantly among nations, and some regions are still experiencing increasing rates.

It is estimated that there will be 8 billion people on Earth in November 2022, and this number is projected to increase to 10.4 billion by 2080, remaining stable until the end of the century. The main reason for the declining population growth rate is lower fertility, despite the fact that average human mortality rates have been decreasing. Reproductive behavior patterns are influenced by factors such as cultural customs, financial status, availability of contraception, and ecological factors like *population density*, which in turn determine fertility.

Keyword

Population density is a measurement of population per unit land area. It is mostly applied to humans, but sometimes to other living organisms too.

The age distribution of the population has changed due to decreasing human mortality in recent decades. According to projections by the Population Division of the UN Department of Economic and Social Affairs, the global number of adults aged 65 and above will double by 2050 compared to children under five. Consequently, there will be fewer individuals of reproductive age in the coming decades, leading to a slowdown in population growth.

Did you get it?

1. **What is the difference between exponential and logistic population growth?**
2. **What is the meaning of logistic population?**

6.3. DEMOGRAPHIC DIVIDEND

Learning Objectives

- Define demographic dividend
- Know the causes of demographic dividend
- Identify different types of demographic dividend
- Focus on labor supply
- Understand human capital

A country's economy can grow as a result of a shift in the age distribution of its population, a phenomenon known as the demographic dividend. Usually, a drop in fertility and mortality rates causes a shift in age structure. Although child survival rates in the majority of countries have improved, many of them—especially the less developed ones—still have high birth rates. Consequently, these nations hardly ever experience the demographic dividend—a financial gain.

A nation that experiences faster economic growth as a result of falling fertility and mortality rates is said to be reaping the demographic dividend. An economy benefits from the rise in working population productivity that results from low birth rates combined with low death rates. The proportion of young dependents in the working population decreases as fewer births are reported. An economy can accelerate its economic development and the prosperity of its citizens by investing its resources in other areas when there are more people in the labor force and fewer people to support.

To reap the benefits of demographic dividends, a nation must undergo a demographic transition, which involves moving from an urban industrial society with low fertility and mortality rates to a predominantly rural agrarian economy with high fertility and mortality rates. During the first part of this shift, fertility rates decline, which causes the labor force to momentarily expand more quickly than the population that depends on it. In addition, per capita income grows faster during this period, all other things being equal. For a nation that has undergone the demographic shift, this economic benefit represents the first dividend.

6.3.1. Causes of Demographic Dividend

Remember

A decline in fertility and mortality rates boosts working population productivity, which leads to a demographic dividend.

Changes in population structure occur due to:

1. Falling birth rate;
2. Lower fertility rate;
3. Increased longevity.

Falling birth rates and lower fertility rates will contribute to a reduction in expenditure, while increased longevity will lead to an increase in the size of the working-age population.

6.3.2. Types of Demographic Dividend

The first dividend period, usually lasting five decades or longer, is quite long. However, the lower birth rate eventually slows the expansion of the labor force. The elderly population is growing due to advancements in medicine and healthier lifestyle choices, which eliminates the demographic dividend and drains additional revenue. When everything else is equal, per capita income grows more slowly at this point, and the first demographic dividend starts to decrease.

An older working population that plans to retire later in life has a strong motivation to build up their financial resources. These assets, typically invested in both domestic and foreign investment vehicles, increase the national income of a nation. The term "second dividend" refers to the rise in national income, which is earned perpetually.

A demographic shift does not always bring benefits, nor is it assured. Any demographic dividend is contingent upon the government's ability to enact the appropriate policies in the domains of governance, the economy, health care, and education. Furthermore, the productivity of young adults determines the demographic dividend a country receives. Productivity is contingent upon several factors, including educational attainment, work policies in a nation, the timing and frequency of childbearing, and economic policies that facilitate the employment of young parents. The productivity of older adults, which is influenced by health care plans, pension and retirement plans, and tax incentives, is also linked to the dividend amount.

There are four main areas where a country can find demographic dividends:

1. Savings—During the demographic period, personal savings grow and can be used to stimulate the economy.

2. Labor supply—More workers are added to the labor force, including more women.
3. Human capital—With fewer births, parents are able to allocate more resources per child, leading to better educational and health outcomes.
4. Economic growth—GDP per capita is increased due to a decrease in the dependency ratio.

Question

Who coined the term Demographic Dividend?

Answer

The term Demographic Dividend was coined by David Bloom. He emphasized the importance of demography to economic growth. David Bloom attributed a large portion of the economic growth of East Asia from 1965 to 1990 to the region's working-age population, which led to an increase in productivity.

Did you get it?

1. **Why is Demographic Dividend Important?**
2. **Which country is in Stage 4 of Demographic Transition?**

6.4. COMPONENTS OF POPULATION CHANGE

Learning Objectives

- Discuss natural increase
- Explain migration
- Focus on population composition
- Understand the impact of population change

Demographers, or researchers who study population change, can now investigate the question of how the United States population changed in the previous year with the release of the 2015 county and metro/micro area population estimates and components of change.

They start by examining the components of change. Births, deaths, and migration are the three elements of change. Natural change or increase refers to the change in population that results from births and deaths combined. Populations expand or contract based on whether they add individuals more quickly than they depart. Examining the distinct blend of migratory patterns and natural change in a region aids in our comprehension of the reasons behind and pace of population shifts.

6.4.1. Natural Increase

The difference between births and deaths in a population is known as natural change. Natural change is often positive, meaning that birth rates are higher than death rates. The term "natural increase" refers to this beneficial natural change. The Salt Lake City metro area in Utah is one example of a naturally occurring increase in the United States.

About 19,100 births and 6,400 deaths occurred in Salt Lake City in 2014 and 2015. With roughly 12,700 more births than deaths, Salt Lake City experienced a natural increase in population of about 12,700, which was a major factor in the city's population growth over the course of the year.

Natural decrease, in which there are more deaths than births, is the opposite of natural increase and can lead to a decline in population. There is frequently a natural decline in aging populations. Maine and West Virginia experienced a natural decline in population between 2014 and 2015. West Virginia and Maine saw 940 and 450 more deaths and births, respectively, between 2014 and 2015. According to the most recent estimates, both cases' populations decreased between 2014 and 2015, partially due to natural decrease.

6.4.2. Migration

Moving from one place to another is called migration. When referring to the difference between the number of people who move into and out of an area, it is commonly expressed as net migration. An area's population is growing when there is a positive net migration. Migration can be divided into two categories: international and domestic.

- *Domestic migration* refers to people moving between areas within a country (here with the example of the United States) and is often one of the largest contributors to population change. Regionally, the South gains the most net domestic migrants, with roughly 440,000 more people moving into southern states than leaving them between 2014 and 2015. Sometimes net domestic migration is negative, in which case more people are moving away than are moving in. The Chicago metro area in Illinois, Indiana, and Wisconsin lost about 80,000 people through migration between 2014 and 2015, which is consistent with a long-standing pattern of negative net domestic migration for the metro area.
- *International migration* refers to people moving into and out of the United States and consists of a diverse group of people such as foreign-born immigrants from many countries around the world, members of the U.S. Armed Forces, and U.S. citizens working abroad. Some areas, like the Miami metro area in Florida, grow (in part) due to net international migration. Miami gained about 70,000 net international migrants between 2014 and 2015, making net international migration a major factor in Miami's population growth.

Did you know?

Fifty years ago, the United States population was a little over 211 million. Currently, the population is almost 335 million. That is an increase of about 2.5 million people per year.

6.4.3. Population Composition

The term "population composition" refers to the attributes of a population, including age, sex, marital status, level of education, type of employment, and relationship to the head of household. The most commonly used of these are the age and *sex composition* of any given population. The population's current and future social and economic circumstances are significantly influenced by the number and proportion of males and females in each age group.

6.4.3.1. Age

One of the fundamental demographic traits that is useful for planning socio-economic development and demographic analysis is the age distribution of a population. Populations in less developed nations tend to be younger, while those in more developed nations tend to be older. The proportions of each age group have a big influence on social and economic policies as well as how people live their lives.

> **Keyword**
>
> **Sex composition** is one of the most basic demographic characteristics of the human population, which is critical for any meaningful demographic analysis.

The ratio of individuals in economically productive age groups to those in dependent age groups is known as the age dependency ratio. The elderly (over 65) and the young (under 15) are the age groups that are dependent. Economically productive people are those who are between the ages of 15 and 64. One common measure of a nation's economic burden is the age dependency ratio. There are a lot of young and old people who depend on the middle-aged population's economic productivity in nations with high age dependency ratios. Because of the high percentage of children in the population, nations with the highest birth rates also have the highest age dependency ratios.

6.4.3.2. Sex Ratio

The number of males for every 100 females in a population is known as the "sex ratio," which represents the male-to-female ratio. Because males and females migrate and die in different ways within the population, sex ratios may differ. The sex ratio in the population, particularly among young adult males, could be impacted by the fact that men are more likely to be mobile, migrate to other regions or countries, and are more susceptible to war.

6.4.3.3. Population Pyramid

The population pyramid combines the two important variables, sex and age. An age and sex composition graph of a population is called a population pyramid. Horizontal bars are used to show the proportions or numbers of men and women in each five-year age group. Because of patterns in migration, birth rates, and death rates from the past and present, population pyramids in different countries can vary significantly (Figure 6.4).

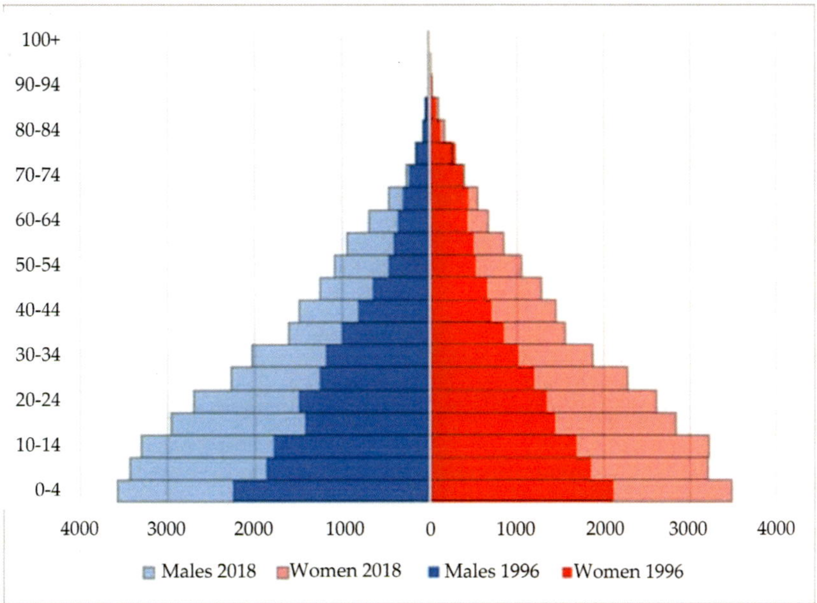

Figure 6.4. Population distribution by age and sex for the Roma population, 1996 and 2018.

Source: Nestorová Dická, J. (2021). Demographic Changes in Slovak Roma Communities in the New Millennium. *Sustainability*. *13*(7), 3735. https://doi.org/10.3390/su13073735.

6.4.3.4. Ethnicity

Acquiring knowledge about ethnicity can aid in the fight against any issues related to inequality. Ethnic data is useful for identifying economic problems, especially in the healthcare sector. This is so that access to health care can be impacted by people's socioeconomic positions, which are influenced by their ethnicity. Racism also has an impact on the standard and accessibility of medical care. Furthermore, it matters because sociopolitical factors are affected by it, especially in highly ethnically diverse countries.

Example

Sri Lanka is very ethnically diverse, with a variety of different ethnic groups. Politics is also ethnically based. This means that decisions regarding political choices are made by people within Sri Lankan society, depending on their ethnic group. Thus highlighting the importance of understanding ethnicity in a population.

6.4.3.5. Other Factors

Because economic, political, and social factors have an impact on population composition, other factors are also important. Education and marital status, for instance, can affect politics and the economy. Density and distribution have an impact on social interactions within a community. Wealth can also affect politics, society, and the economy. Each of these population composition factors accomplishes this in a unique way. Governments can, however, plan how to ensure that resources, infrastructure, and services are distributed fairly among a population by gathering data on population composition. In addition, population trends are being watched to determine whether or not population control measures are necessary. This measure would be executed if a population is deemed excessively young or elderly, thereby affecting the nation's economy.

Remember

Age, sex, and ethnicity are the three major elements of population composition.

Question

What is an example of population change?

Answer

In 2021, California's census revealed that the state's population decreased by 173,000 since the last census, marking the first time in the state's history that the population didn't grow. This is most likely the result of emigration in response to natural disaster in the form of wildfires in addition to a lower natural population change due to deaths from the COVID-19 pandemic.

6.4.4. Impact of Population Change

Human population growth impacts the Earth system in a variety of ways, including:

- Increasing the extraction of resources from the environment. These resources include fossil fuels (oil, gas, and coal), minerals, trees, water, and wildlife, especially in the oceans. The process of removing resources, in turn, often releases pollutants and waste that reduce air and water quality, and harm the health of humans and other species.
- Increasing the burning of fossil fuels for energy to generate electricity, and to power transportation (for example, cars and planes) and industrial processes.
- Increase in freshwater use for drinking, agriculture, recreation, and industrial processes. Freshwater is extracted from lakes, rivers, the ground, and man-made reservoirs.
- Increasing ecological impacts on environments. Forests and other habitats are disturbed or destroyed to construct urban areas including the construction of homes, businesses, and roads to accommodate growing populations. Additionally, as populations increase, more land is used for agricultural activities to grow crops and support livestock. This, in turn,

can decrease species populations, geographic ranges, biodiversity, and alter interactions among organisms.

- Increasing fishing and hunting, which reduces species populations of the exploited species. Fishing and hunting can also indirectly increase the numbers of species that are not fished or hunted if more resources become available for the species that remain in the ecosystem.
- Increasing the transport of invasive species, either intentionally or by accident, as people travel and import and export supplies. Urbanization also creates disturbed environments where invasive species often thrive and outcompete native species. For example, many invasive plant species thrive along strips of land next to roads and highways.
- The transmission of diseases. Humans living in densely populated areas can rapidly spread diseases within and among populations. Additionally, because transportation has become easier and more frequent, diseases can spread quickly to new regions.

Did you get it?

1. **What are the three main changes in population?**
2. **What are the factors for population change?**

ROLE MODEL

THOMAS ROBERT MALTHUS

English Economist and Demographer

Malthus was interested in everything about populations. He accumulated figures on births, deaths, age of marriage and childbearing, and economic factors contributing to longevity. His main contribution was to highlight the relationship between food supply and population. Humans do not overpopulate to the point of starvation, he contended, only *because* people change their behavior in the face of economic incentives.

Noting that while food production tends to increase arithmetically, population tends to increase naturally at a (faster) geometric rate, Malthus argued that it is no surprise that people thus *choose* to reduce (or "check") population growth. People can increase food production, Malthus thought, only by slow, difficult methods such as reclaiming unused land or intensive farming; but they can check population growth more effectively by marrying late, using contraceptives, emigrating, or, in more extreme circumstances, resorting to reduced health care, tolerating vicious social diseases or impoverished living conditions, warfare, or even infanticide. Malthus was fascinated not with the inevitability of human demise, but with why humans do *not* die off in the face of such overwhelming odds. As an economist, he studied responses to incentives.

Malthus is arguably the most misunderstood and misrepresented economist of all time. The adjective "Malthusian" is used today to describe a pessimistic prediction of the lock-step demise of humanity doomed to starvation via overpopulation. When his hypothesis was first stated in his best-selling *An Essay on the Principle of Population* (1798), the uproar it caused among noneconomists overshadowed the instant respect it inspired among his fellow economists. So irrefutable and simple was his illustrative side-by-side comparison of an arithmetic and a geometric series—food increases more slowly than population—that it was often taken out of context and highlighted as his main observation. The observation is, indeed, so stark that it is still easy to lose sight of Malthus's actual conclusion: that because humans have *not* all starved, economic choices must be at work, and it is the job of an economist to study those choices.

Malthus addressed many other issues. His *Principles of Political Economy* (1820) was the first text to describe a demand schedule as separate from the quantity demanded at a given price. His exposition of demand curves clarified the debate on Say's law and gluts (to which he objected in the long run on the grounds that markets self-adjust). His work centered on contrasting the long run, as exemplified by population growth, with the short run, reflected by cyclical events such as those affecting agriculture. Writing before the Industrial Revolution, Malthus did not fully appreciate the impact of technology (i.e., pesticides, refrigeration, mechanized farm equipment, and increased crop yields) on food production.

Malthus died in 1834, before seeing economics characterized as the "dismal science." That phrase, coined by Thomas Carlyle in 1849 to demean John Stuart Mill, is often erroneously thought to refer to Malthus's contributions to the economics of population growth.

CASE STUDY

RAPID POPULATION GROWTH – NIGER

The majority of Niger's population are subsistence farmers. This results in a high birth rate because people look to have large families to increase their economic opportunities. Which coupled with the high infant mortality rate means that many families opt to have more children than they need to ensure that enough survive to work on the land.

The birth rate is also high in Niger due to the cultural beliefs of the Hausa ethnic group, where a large family is viewed as a status symbol. Within the community, it is acceptable for women to marry at the age of 14, meaning there is a greater chance of them having a large family.

Within Niger there is limited access to contraceptives and many of the Hausa women are reluctant to use them as their husbands forbid them, all of which leads to a high birth rate.

FOCUS ON CAREERS

DEMOGRAPHER

A demographer is a professional who studies statistical information about a population. This includes studying items like birth rates, death rates, age ratios, ethnicities and living conditions. Demographers can work for various institutions, such as government agencies, universities and political organizations.

What does a demographer do?

Demographers collect information about a population through various statistical tools, such as surveys. Afterward, they compile all results and evaluate them. By conducting an advanced statistical analysis, demographers make predictions or provide valuable insights about that population. Demographers may spend their time working in the field, such as going door-to-door to collect survey responses. They can compile and analyze data in a research setting.

Skills of a Demographer

Demographers rely on a range of skills to collect and analyze data, including:

Attention to Detail

Demographers rely on gathered data to perform statistical analyzes. It's essential they have accurate information and that they can perform correct mathematical operations with this data. Therefore, demographers often require a strong attention to detail, as inaccurate findings can have a drastic impact on their final analysis. Demographers also know how to identify issues in a data set and can use this skill to check the accuracy of their calculations routinely.

Communication

Demographers rely on communication skills to connect with several individuals. For example, they often interact with the people they survey. This may involve going door-to-door to interview people or conducting a questionnaire over the phone. Demographers need strong communication skills to ask their questions effectively and record the answers, while also offering a pleasant experience for the individual.

Demographers also interact with other professionals in the workplace. They may conduct their study alongside other demographers or present their findings to colleagues and supervisors within an organization. Demographers use their communication skills to work well with others and accurately describe their findings, both verbally and in writing.

Computer Literacy

To perform data collection and statistical analysis processes, demographers typically rely on computer software. This software allows them to input large quantities of data and quickly perform various computations. Demographers can also use this software to create informative charts and graphs, which help illustrate their statistical findings. Therefore, demographers typically have a high comfort level with computer hardware and software, as they often use technology for day-to-day work tasks.

Math Skills

Demographers use basic math techniques, such as addition and percentage calculations, to calculate data. Demographers also use advanced mathematical techniques to analyze datasets and identify research results. For example, they might determine if a result is statistically relevant or whether an event has a high chance of occurring. In many situations, demographers work with complex data sets, so it's important to know the correct mathematical operations to use to draw meaningful results.

Research Skills

Research skills involve the ability to gather information in an efficient and accurate manner. Demographers use their research skills to learn more about their audience and design study parameters. For example, a demographer may research a neighborhood to determine the best way to interview the constituents there. Demographers also use their research skills to discover similar studies that can inform their own statistical analysis and provide more support for their findings.

Time Management

It's important for demographers to conduct research and perform analysis efficiently. Organizations and clients often require information from demographers before a certain date so they can make timely decisions. For example, a political campaign may hire a demographer to provide population information that can help determine where to run television ads. Demographers can benefit from the ability to plan the timeline of a survey process, including how long it might take to analyze results and deliver a report before an assigned deadline.

How to Become a Demographer?

Follow the steps below to become a demographer:

Earn a Bachelor's Degree

To begin a demography career, it's important to earn a relevant bachelor's degree. Common areas of study for bachelor's degree programs include sociology, statistics and economics. During these programs, students learn the basic math principles they need to perform statistical studies, alongside the proper procedures for conducting this type of research. Students also learn about the different fields that use demographic studies and how their statistical results might influence decisions within that field.

Gain Relevant Experience

Many organizations that hire demographers look for candidates with related work experience. For your first position as a demographer, you can earn related work experience in a few ways. One example is working as a research assistant during your undergraduate years by helping a university professor with research projects involving population statistics. As a research assistant, you may help collect data or prepare reports, which can provide critical skills for an entry-level demographer position. Another example is completing an internship or work-study program. These programs allow you to gain real-life experience in your desired field while completing your degree. Many undergraduate programs require an internship or work-study program, while others provide key resources to help you find one. You can also look for part-time or volunteer positions where you can gain the skills required of a demographer. For example, you can volunteer for a political campaign or find a job that involves using statistical analysis software.

Get an Advanced Degree

A common requirement for demographers is a master's degree. During a master's degree program, students learn more about how to perform advanced statistical analyzes and how to apply them to different fields. Students also can work more closely with experienced demographers in their desired industry. Common areas of study for master's degree students include sociology and statistics.

Create a Resume

To apply for a demographer position, candidates typically need a resume. This resume contains information about your undergraduate and master's degree studies, including the courses you took and any special achievements. Demographer resumes also contain information about related work experience, such as time spent as a research assistant or administering surveys. Spend time crafting a professional one-page resume that's tailored for a specific demographer position.

Apply for Demographer Jobs

Research online job boards for entry-level demographer positions, including temporary, part-time or full-time assignments. Depending on your circumstances, you may benefit from connecting with a professor at your previous institution for job search advice or networking opportunities. You can also connect with a mid-level or senior-level demographer in your preferred industry for an informational interview, as they may have some keen insights into the job application process.

A CLOSER LOOK

GLOBAL AGING

Population aging is the dominant demographic trend of the twenty-first century—a reflection of increasing longevity, declining fertility, and the progression of large cohorts to older ages. Never before have such large numbers of people reached ages 65+ (the conventional old-age threshold). We expect to add 1 billion older individuals in the next three to four decades, atop the more than 700 million older people we have today. Among the older population, the group aged 85+ is growing especially fast and is projected to surpass half a billion in the next 80 years. This trend is significant because the needs and capacities of the 85+ crowd tend to differ significantly from those of 65-to-84-year-olds.

Although every country in the world will experience population aging, differences in the progression of this phenomenon will be considerable. Japan is currently the world leader, with 28% of its population 65 and over, triple the world average. By 2050, 29 countries and territories will have larger elder shares than Japan has today. In fact, the elderly numbers in the Republic of Korea will eventually overtake Japan's, reaching the historically unprecedented level of 38.1%. The median age (48.4) in Japan is also currently the highest of any country and more than twice that of Africa (19.7). But by 2050, Korea (median age 56.5 in 2050) is also expected to overtake Japan on that metric (54.7).

Three decades ago, the world was populated by more than three times as many adolescents and young adults (15- to 24-year-olds) as older people. Three decades from now, those age groups will be roughly on par.

By income group, the sharpest growth in the numbers of older people will occur in countries currently classified as middle-income. This is unsurprising, as these countries make up 74% of the global population. What may be surprising is that the older population share in middle-income countries is increasing at a much faster rate than in their low- and high-income counterparts. Moreover, in comparison with high-income countries, today's middle-income countries are projected to have appreciably greater real incomes when their older population shares reach comparably elevated levels. This contradicts the common claim that developing economies are getting old before they get rich.

The challenge middle-income countries face is not predominantly insufficient income to take care of their older people. Rather, it is how well institutions and policies can promote economic and social security among older people in a financially sustainable way.

Population aging is sounding alarms worldwide. Whether increased longevity is associated with more or less of a person's life lived in frailty is among the most salient unresolved questions public and private policymakers throughout the world face.

Economists continue to express concerns. These relate to downward pressure on economic growth due to labor and capital shortages and falling asset prices in the future as a growing and more aged cohort of older people seeks to support itself by liquidating investments. Another major issue has to do with fiscal stress. Government coffers will be strained by rising pension liabilities and the cost of health and long-term care associated with the expected growth in the incidence and prevalence of chronic diseases such as cancer, among others. These challenges will, however, be partially offset by the increasing, but typically neglected, value older people create through productive nonmarket activities like volunteer work and caregiving.

Without historical lessons from a world with such large numbers of older people, there is even more uncertainty about our collective future. However, adopting a business-as-usual approach to the challenges of population aging would be irresponsible.

Various responses could cushion the economic burden of population aging. These include policy reforms to promote the financial sustainability and intergenerational equity of health and pension financing. Raising the legal age of retirement, which has been relatively stable in nearly all countries for the past several decades would also ease the burden. Pronatalist tax incentives are also a policy option for the long term, but their effect on fertility is thus far unproven.

Additional approaches include efforts to increase the emphasis on early detection and prevention of disease through, for example, better awareness of the benefits of physical activity and subsidization of such activity. Relaxing the institutional and economic barriers to international immigration from regions with relatively large working-age populations could alleviate labor shortages.

Finally, technological innovations are likely to ameliorate the effects of population aging. New drugs to slow the process of aging and add healthy years to people's lives and the invention and deployment of assistive devices such as robots are two among many such improvements. Institutional innovations like new models of home health care, public transportation systems, the design of urban layouts, and financial instruments are also on the horizon.

SUMMARY

- The workforce of a company is collectively known as human resources (HR). It can also refer to the division overseeing the workforce. Human resources management, also called HR or personnel department, falls under a dedicated HR department in many organizations.
- The specific duties of HR vary depending on the organization's size, industry, and structure.
- Strategic initiatives began to be implemented in HR departments from the 1980s onwards. This movement was based on research into how employee-related issues impact a company's long-term financial performance.
- Companies can choose between in-house HR departments or third-party companies for HR needs. Internal HR management is in-house, while external HR management is outsourced.
- Population growth is the change in the number of individuals of a species in a specific area over time in population ecology.
- Sociologists view population issues as crucial for understanding human interactions. The four population theories discussed are: Malthusian, zero population growth, cornucopian, and demographic transition theories.
- Population loss, from mortality and emigration rates, can be subtracted from population gain, from fertility and immigration rates, to calculate population growth rates in a specific area and time period.
- The natural environment is crucial for population growth due to its carrying capacity. Limited life-sustaining resources can be reduced for individual species members by factors like increased population density and competition from other species.
- The global population has grown exponentially since the 18th century, but growth rates have been decreasing since the second half of the 20th century. Although population growth rates vary significantly among nations, overall growth is declining, with some regions still experiencing rising rates.
- A country's economy can grow as a result of a shift in the age distribution of its population, a phenomenon known as the demographic dividend.
- Falling birth rate and lower fertility rate will contribute to a reduction in expenditure, while the increased longevity will lead to an increase in the size of the working-age population.
- Any demographic dividend is contingent upon the government's ability to enact the appropriate policies in the domains of governance, the economy, health care, and education.
- Births, deaths, and migration are the three elements of change. Natural change or increase refers to the change in population that results from births and deaths combined. Populations expand or contract based on whether they add individuals more quickly than they depart.

- Moving from one place to another is called migration.
- The term "population composition" refers to the attributes of a population, including age, sex, marital status, level of education, type of employment, and relationship to the head of the household.

MULTIPLE CHOICE QUESTIONS

1. **What does carrying capacity refer to?**
   a. The ability of a community to welcome new immigrants
   b. The capacity for globalism within a given ethnic group
   c. The amount of life that can be supported sustainably in a particular environment
   d. The amount of weight that urban centers can bear if vertical growth is mandated
2. **What three factors did Malthus believe would limit human population?**
   a. Self-preservation, old age, and illness
   b. Natural cycles, illness, and immigration
   c. Violence, new diseases, and old age
   d. War, famine, and disease
3. **What does cornucopian theory believe?**
   a. That human ingenuity will solve any issues that overpopulation creates
   b. That new diseases will always keep populations stable
   c. That the earth will naturally provide enough for whatever number of humans exist
   d. That the greatest risk is population reduction, not population growth
4. **Which of the following processes is responsible for fluctuation in population density?**
   a. Mimicry
   b. Natality
   c. Hibernation
   d. Age structure
5. **Which letter is used to designate the birth rate (Natality)?**
   a. B
   b. E
   c. I
   d. K
6. **Which of the following processes is responsible for fluctuation in population density?**
   a. Aestivation
   b. Camouflage

c. Hibernation

d. Mortality

7. Which of the following is responsible for an increase in population density?

a. Age structure

b. Emigration is more

c. Immigration is more

d. Death rate increase

8. Which letter is used to designate the death rate (Mortality)?

a. J

b. D

c. S

d. U

9. Which of the following processes is responsible for fluctuation in population density?

a. Hibernation

b. Aestivation

c. Mimicry

d. Immigration

10. Which of the following is responsible for an increase in population density?

a. Death rate increase

b. Age structure

c. Birth rate increase

d. Emigration is more

Review Questions

1. What Is the role of human resources?
2. What are the types of human resources?
3. What causes population growth?
4. What is the determination of population growth?
5. What are the components of the population change equation?
6. How are births and deaths measured?

Answer to Multiple Choice Questions

1. (c)	2. (d)	3. (a)	4. (b)	5. (a)
6. (d)	7. (c)	8. (b)	9. (d)	10. (c)

REFERENCES

1. Andregg, M., (2013). *Seven Billion and Counting: The Crisis in Global Population Growth*. Minneapolis: Twenty-First Century Books.
2. Annan, K., Christopher, F., Linda, S., (2002). *State of the World 2002: A Worldwatch Institute Report on Progress Towards a Sustainable Society*. New York: W.W. Norton.
3. Gomes, C., Lily, K., & Orlando, W., (2020). *Religion, Hypermobility and Digital Media in Global Asia*. NL Amsterdam: Amsterdam University Press. http://dx.doi.org/10.5117/9789463728935.
4. Liotta, P. H., (2012). *The Real Population Bomb: Megacities, Global Security & the Map of the Future*. Washington, D.C: Potomac Books.
5. Robertson, T., (2012). *The Malthusian Moment: Global Population Growth and the Birth of American Environmentalism*. New Brunswick: Rutgers University Press.
6. Whitmore, H. W., (2007). *The World Economy, Population Growth, and the Global Ecosystem*. New York: Palgrave Macmillan US. http://dx.doi.org/10.1057/9780230607309.

CHAPTER 7

Settlements and Urbanization

LEARNING OBJECTIVES

After studying this chapter, you will be able to:

- Discuss rural settlements;
- Focus on urban settlements.

"Urbanization is not about simply increasing the number of urban residents or expanding the area of cities. More importantly, it's about a complete change from rural to urban style in terms of industry structure, employment, living environment, and social security."

—*Li Keqiang*

INTRODUCTORY EXAMPLE

Rural Tourism

Traveling outside of major cities and well-known tourist destinations is known as rural tourism. Traveling to rural areas, such as villages, cottages, homestays, farms, ranches, or eco lodges, involves visiting undeveloped, sparsely populated areas that are frequently dependent on agriculture. Engaging in outdoor sports, hiking, camping, and spending time in nature are all viable activities when traveling in rural areas. It has to do with going off the usual path, engaging in outdoor activities and sports, spending time in nature, and ethical and sustainable tourism. It has a significant chance of improving the experience and making travel more responsible. Travelers who visit rural locations have the opportunity to fully experience local culture, avoid tourist traps and crowds, and appreciate the advantages of being outside. Rural tourism is very important in developing nations. Families in remote, rural areas that aren't often visited by tourists benefit directly from it. Additionally, it offers chances for growth. It enables people in developed nations to unwind and take a break from their hectic lives. Reduced negative effects from excessive tourism are beneficial in any country.

Rural tourism comes in a variety of shapes and forms, some of which include:

- Agritourism might bring tourists to the area and give farmers new sources of income, which would be economically beneficial for rural areas. Additionally, it can support the advancement of sustainable agricultural methods and public awareness of the importance of agriculture to local economies and food security.
- Ecotourism is a type of tourism development that encourages responsible travel to environmentally sensitive areas with the goals of preserving the environment, assisting the local population, and educating visitors about the natural world. Ecotourism developments are often characterized by outdoor activities and experiences that allow visitors to observe and appreciate nature while reducing their impact on it.
- The tourism sector may help advance understanding of cultural appreciation, the preservation of historic sites, and the economic growth of nearby communities. Since many cultural tourism experiences are based on responsible travel ideas like respect for local culture and the environment, as well as support for communities, it may also help promote sustainable tourism practices.
- Nature tourism allows tourists to discover and appreciate the beauty of natural ecosystems such as national parks, forests, rivers, mountains, and coastal locations. Many nature tourism activities are based on sustainable tourism principles and aim to mitigate the environmental impact of tourists while providing educational and learning experiences.

Depending on the resources and attractions in the region, these types of rural tourism can be combined or offered separately.

INTRODUCTION

A sort of human settlement found throughout a big city is called an urban settlement. Stated differently, it refers to the act of adapting and becoming a part of a previously deserted setting. Communities of homes, which can be villages or towns, are home to people. The term "urban migration" describes how people move from rural to urban areas, how the population of rural areas declines in tandem with this movement, and how local communities adjust to these changes.

7.1. RURAL SETTLEMENTS

Learning Objectives

- Identify the main factors influencing the rural settlements
- Explain rural settlement types
- Understand patterns of rural settlement
- Focus on problems of rural settlement

Rural communities are constructed in close proximity to their land. Primary industries like farming, fishing, and animal husbandry dominate them. Rural settlements are relatively small in size. The residents of rural settlements rely on small-scale mining, quarrying, forestry, and soil exploitation for their livelihood. The primary group of farmers and farm laborers in a typical village is supported by secondary workers such as retailers, educators, clergy, publicans, postmasters, blacksmiths, and garage owners. Additionally, with urbanism quickly becoming the new way of life, villages consist of a mix of younger residents who reside in the village and older residents who commute to work in nearby towns. The percentage of people in each of these classes contributes to the overall population of the village, which varies depending on the type of farming practiced, *soil quality*, site accessibility and appeal, and location within the settlement pattern.

The main factors influencing the rural settlements are:

1. Nature of the topography
2. Local weather conditions
3. Quality of the soil
4. Nature of surface and subsurface water
5. Pattern of landholding
6. Social organization

Rural settlements with only a few houses, true village communities, villages, and large villages are all considered hamlets, depending on their size. Genuine village communities are simply old, established settlements where, in the worst circumstances, there is no such thing as private property and everything is owned by the peasant community. These kinds of village communities can be found in France, Malaya, and India.

In addition to farming villages, there are industrial villages, fishing villages, forest villages, mining and quarrying villages, tourist industry-dependent villages, and dormitory villages that serve neighboring towns.

Each has its own characters :

- Many industrial and mining villages are unprepossessing, suffering from ugly and drab surroundings.
- Fishing and tourist villages are more attractively situated and, in many cases, possess the charm and interest of historical buildings.
- The majority of dorm villages are either brand-new or mostly made up of contemporary housing estates. All of these settlements have far more specialized jobs than towns do, and common urban functions like wholesaling and administration are hardly ever seen. A nation's total number of villages and their significance in the social and economic spheres are significant factors. For instance, in India, over three-quarters of the *population* resides in more than 600,000 villages, the majority of which are home to fewer than 500 people. Worldwide, an estimated two out of every three people still reside in villages, hamlets, and dispersed homes.

Keyword

Soil quality is how well soil does what we want it to do. Soil quality is the capacity of a specific kind of soil to function to sustain plant and animal productivity, maintain or enhance water and air quality, and support human health and habitation.

7.1.1. Rural Settlements Types

There are four different kinds of rural settlements: hamleted, semi-compact, compact, and dispersed or scattered (Figure 7.1).

Figure 7.1. A rural settlement in Pahang, Malaysia.

Source: By Chongkian – Own work, CC BY-SA 4.0, https://commons.wikimedia.org/w/index.php?curid=47812778.

Keyword

Population is defined as all nationals present in, or temporarily absent from a country, and foreigners permanently settled in a country.

7.1.1.1. Dispersed Settlements

Houses that are widely separated or dispersed over a wide area are characteristics of dispersed settlements. Small housing clusters with low population densities may be found in these settlements. The houses are frequently surrounded by sizable tracts of agricultural land. These communities are usually found in areas rich in natural resources and are frequently connected to agricultural pursuits like small-scale farming, ranching, and livestock farming. There are numerous ways to create dispersed settlements, including farmsteads, homesteads, and ranches (Figure 7.2).

Keyword

Colonization is a process of establishing control over specific territories or peoples for the purpose of cultivation, possibly involving settling, the establishment of colonialism, coloniality, and colonies.

Figure 7.2. Dispersed settlement or *Streusiedlung* in Brülisau, AppenzellInnerrhoden, Switzerland.

Source: By Roland Zumbühl (Picswiss), Arlesheim (Commons:Picswiss project), CC BY-SA 3.0, https://commons.wikimedia.org/w/index.php?curid=1192725.

7.1.1.2. Farmsteads and Homesteads

Each household in these dispersed settlements is situated on a separate farm or agricultural estate. They are typically found in regions where agriculture serves as the main source of income. Large landholdings and a self-sufficient lifestyle are common characteristics of farmsteads and homesteads. Farmsteads are typical in regions like rural Europe and the Midwest of the United States.

7.1.1.3. Ranches

Large agricultural estates known as ranches are reliant on the farming of livestock, such as sheep or cattle. These communities are frequently located in isolated locations with sizable grazing lands. In Australia, Canada, and the western United States, ranches are widespread.

7.1.1.4. Nucleated Settlements

Nucleated settlements are typified by a collection of homes in a specific area, frequently centered around a marketplace or hub. They are typically found in regions with rich soil, where the primary industry is agriculture. Nucleated settlements are distinguished by common facilities like a church or temple and frequently have a strong sense of community. These communities can be found in regions like Europe and some parts of Asia where there has been extensive *colonization*. These communities can be further separated into two categories: dispersed nucleated communities and compact nucleated communities. Another name for nucleated settlements is clustered settlements. Villages and hamlets are just two examples of the various shapes that clustered settlements can take.

7.1.1.5. Villages and Hamlets

The population of these small settlements ranges from a few hundred to a few dozen households. They are frequently found in places where people practice animal husbandry or subsistence farming due to low agricultural productivity. Basic facilities like a primary school, a small clinic, and a few stores are typically found in hamlets and villages.

Question

What is meant by rural settlement?

Answer

Rural settlements are built closely and directly related to land. They are dominated by primary activities such as fishing, animal husbandry, agriculture, etc. The size of rural settlements is comparatively small.

7.1.2. Patterns of Rural Settlement

A building's or house's relationship to another is known as the pattern of settlement. It can be recognized by looking at and reading a map at the local scale. Only compact and semi-compact settlement patterns are covered by the pattern; dispersed settlement takes on a unique shape. Its size and form are strongly influenced by sociocultural elements like caste systems and human needs. The following patterns are used to categorize the rural settlements: Nebular, Star-like, Triangular, Rectangular, Linear, Circular, and Semicircular. In valleys and mountainous regions, the settlement is linear; in fertile plains, it is rectangular; around lakes and ponds, it is circular; at crossroads, it is triangular; in rare instances, it resembles the nebular form; and on river terraces, it is star-shaped.

7.1.2.1. Rectangular Pattern

In rural settlements, this is the most prevalent pattern that is seen. Large valleys and level, fertile alluvial plains are the development sites for rectangular settlements. In rectangular settlements, the streets are straight and perpendicular to one another. Villages in the Sutlej-Ganga Plain and planned communities in Germany, Malaysia, Israel, France, and other countries are instances of this.

7.1.2.2. Linear Pattern

The houses are arranged in a row beside a river, canal, road, or railroad track. The linear pattern results from the physical limitations connected to these locations. Examples of it can be found in the valleys of the Alps and the Rocky Mountains, as well as along the Ganga-Yamuna Plain's roads (Figure 7.3).

Keyword

Lodging refers to the use of a short-term dwelling, usually by renting the living space or sometimes through some other arrangement.

Figure 7.3. Some communities along the Saint Lawrence River in Quebec, Canada, developed as linear settlements, as is still clearly seen in Champlain.

Source: By Rene Beaudoin – Own work, CC BY-SA 3.0, https://commons.wikimedia.org/w/index.php?curid=3835633.

7.1.2.3. Circular and Semi-circular Pattern

Communities that are built next to lakes, on top of mountains, beside meandering rivers, and along seacoasts tend to take on circular or semicircular patterns. The majority of the population works in fishing, salt production, and tourism-related industries like boating, *lodging*, and boarding.

7.1.2.4. Star-Shaped Pattern

Settlements that resemble stars or radials emerge at the intersection of roads. The newly developed areas of the settlement expand outward along the roads. Examples of this include the Yangtze River plains, Punjab, India, and some regions of Northwest Europe.

7.1.2.5. Triangular Pattern

Points like the confluence of rivers give rise to a pattern of settlements that is triangle-shaped.

7.1.2.6. Nebular Pattern

It is a ring-shaped pattern that is growing all around a hub. This center might be a temple, a landlord's residence, or anything else entirely. The size, pattern, and shape of rural settlements can be used to categorize them.

7.1.3. Problems of Rural Settlement

Rural settlements in underdeveloped countries are poorly equipped with infrastructure and are numerous.

- Water Supply: Residents of villages, particularly those in arid and mountainous regions, must travel long distances on foot to obtain drinking water. Waterborne diseases like jaundice and cholera are prevalent in rural areas.
- Drought: South Asian countries frequently experience drought conditions in their regions. The lack of irrigation leads to disruptions in crop patterns.
- Sanitation: The absence of amenities such as waste disposal and toilets results in health issues in these regions.
- Housing: Dwellings constructed from thatch, wood, and mud are vulnerable to damage from floods and heavy rains. They also require regular maintenance.
- Roads: Inadequate infrastructure like modern communication networks and unpaved roads leads to challenges. During the rainy season, these settlements struggle to provide essential services.
- Health and Education: Insufficient educational and healthcare facilities for large rural populations are significant concerns in underdeveloped countries.

Rural living takes many different forms and follows ever-changing patterns. As a result of changes in vocation, types of cultivation, cultural assimilation, technological advancements, and innovation dissemination, kinds and patterns observe the architectural and geometrical alteration and shifting paradigm of society and cultural way of life.

Did you get it?

1. **What are the characteristics of rural settlement?**
2. **What are the three patterns of rural settlement?**

Did you know?

Agriculture is the primary industry in most rural areas. Most people live or work on farms or ranches. Hamlets, villages, towns, and other small settlements are in or surrounded by rural areas.

Did you know?

In the United States, the Census Bureau classifies a rural area as a town with fewer than 1,000 people per 2.6 square kilometers (square mile), and surrounding areas with fewer than 500 people per 2.6 square kilometers (square mile).

7.2. URBAN SETTLEMENTS

Learning Objectives

- Discuss about settlement in urban areas
- Know the classification factors of urban settlement
- Identify types of urban settlements
- Explain three functions of urban settlement
- Understand change in urban functions
- Give examples of urban functions
- Focus on urbanization problems in developing countries

A densely populated area where many people live and work in urbanized surroundings is called an urban settlement. Towns and cities are preferred for urban settlements, regardless of size. Typically, a regional plan partitions cities into zones subject to standards on particular aspects like housing, sanitation, and the environment. The area includes residential areas, infrastructure, services, and spaces for commercial, industrial, and administrative activities.

7.2.1. Settlement in Urban Areas

Every country has a different definition of urban areas (Figure 7.4). The phrase is frequently used to describe the temporary or sporadically permanent way of life that displaced families lead when discussing displaced populations. A regional plan usually divides cities into regulated zones according to requirements for particular industries, like environment, habitat, housing, and *sanitation*.

Figure 7.4. London, a city in the United Kingdom, is a large settlement with a human population of 14 million in its metropolitan area.

Source: By © User: Colin / Wikimedia Commons, CC BY-SA 4.0, https://commons.wikimedia.org/w/index.php?curid=35768815

7.2.2. Classification Factors of Urban Settlement

Typical classification factors include things like administrative structure, occupation structure, and population size.

Population size:

- In Colombia, a town must have at least 1,500 residents in order to be classified as urban; this number is 2,000 in Argentina and Portugal; 2,500 in the United States and Thailand; 5,000 in India; and 30,000 in Japan
- Along with the population number, factors such as the density of 400 people per square kilometer and the percentage of non-agricultural workers are taken into account.

Occupational structure:

- In some countries such as India, the main economic activities are considered to designate a settlement as urban.

Administration:

- A settlement may be classified as urban in various nations based on its administrative structure.
- For instance, a settlement of any size is considered urban in India if it has a municipality, board, or notified area board.

Keyword

Sanitation refers to public health conditions related to clean drinking water and treatment and disposal of human excreta and sewage.

7.2.3. Types of Urban Settlements

Urban areas are classified as cities based on their size, services offered, and functions performed:

- Town: There isn't much to separate a city from a village. The key element is access to resources. In towns, the construction and infrastructure are slightly better. Compared to villages, basic amenities are offered more often in towns.
- Cities: Cities are quite large and have more people than towns. They serve bigger economic purposes. There are more developed systems for transportation, quality of life, education, etc.
- Conurbation: The term "metropolis" refers to extensive merged urban settlements. Initially, these towns or cities were divided. Cities such as Tokyo and Greater London are examples of conurbations.

- Megalopolis: Means "great city." A megalopolis is a very dense cluster of many cities. Another name is Hyper Metropolitan. A classic example is an urban area that stretches from Boston in the north to Washington in the south of the United States.
- Million City: A city is said to be a million city when its population exceeds one million. There are now one million cities worldwide, which is a record number. London became the first city with one million inhabitants in 1800, followed by Paris in 1850, New York in 1860, and more than 80 such cities by 1950 (Figure 7.5).

Figure 7.5. Urban structure of Sortavala, a town in the Republic of Karelia, Russia.

Source: By Abel111222 – Captured by myself, CC BY-SA 4.0, https://commons.wikimedia.org/w/index.php?curid=109417174.

Question

What is urban settlement?

Answer

In essence, an urban settlement is a place where a large number of people live and work. These crowded areas are seen as the antithesis of rural settlements. Every country defines an urban settlement area differently. Common classification factors include organizational structure, occupation structure, and population size. The region also has places for commercial, industrial, and administrative activities, as well as residential areas, services, and infrastructure.

7.2.4. Three Functions of Urban Settlement

Let's examine a few of these urban functions. The functions of industry, places of worship, and residential areas are three distinct examples of the functions of urban settlement.

- **Functions of Industry:** The most prevalent economic activities in urban areas are tertiary and quaternary ones. These consist of the trade, commerce, and service sectors like travel and teaching.

- **Functions of Places of Worship:** Many times, people have very specific uses for places of worship. They foster a sense of community in a place by enabling people to come together and share their faith.
- **Functions of Residential Areas:** Urban areas now host more than half of the world's population. Accordingly, homes are a vital service that many people receive from urban settlements. Numerous people have moved to cities as a result of the expansion of services offered by urban settlements. A need for housing is thus created.

Remember

Influences are changing urban functions on many scales such as regional, national, international and global.

7.2.5. Change in Urban Functions

A place's historical and contemporary ties to other locations impact its social and economic features.

7.2.5.1. Urban Functions: Regional and National Influences

Cities in the UK are divided from north to south; in the south, new development has been prioritized, while in the north, city populations have decreased. Those cities, which were formerly well-known for their heavy industry, have seen a decline in employment due to technology and *globalization*. A government initiative to establish a northern powerhouse in 2015 aimed to facilitate trade and industry connections throughout the nation. A similar emphasis on enterprise by the Conservative government in the 1980s resulted in financial growth and free-market privatization.

7.2.5.2. Urban Functions: International and Global Influences

The UK is a diverse country because of colonization and international ties. Many people, particularly in London, identify as belonging to numerous ethnic backgrounds. Despite the fact that immigration has benefited many towns, prejudice still exists towards ethnic groups. Migrants are relied upon as a younger population with higher aspirations, given the aging population and diminishing skills.

The British government encouraged a large number of African-Caribbean immigrants to come to the UK in order to find employment, a group known as the Windrush Generation. They were entitled to work in the UK without a document because they were British subjects. Despite being a member of the Commonwealth, many immigrants experienced racism, prejudice, and unjust deportation.

Keyword

Globalization is a term used to describe how trade and technology have made the world into a more connected and interdependent place.

7.2.5.3. Demographic Changes

Identity gains significance when a nation undergoes rapid changes. Cities like London blend influences from minorities with identity politics. Southall in West London is renowned for its substantial Somali and Indian communities. A significant portion of the Somali diaspora works to provide financial support to their families and communities back home. Identity politics encompasses not only civil rights struggles and LGBTQ+ social movements but also racial tensions. Canal Street in Manchester emerged as the hub of the "gay village" due to laws that targeted civil liberties, leading the LGBT community to gather for support. This community continues to be served and supported by this area. Lived experiences shaped by interactions with the location itself have the potential to impact a place's engagement and attachment.

Keyword

Flooding is a temporary overflow of water onto land that is normally dry. Floods can result from rain, snow, coastal storms, storm surges, overflows of rivers, and dam failure.

7.2.6. Examples of Urban Functions

Let's consider Berbera as an illustration of an urban function. Berbera is a city in the self-declared autonomous region of Somaliland, located in northwest Somalia. Its seaport, near Aden and along an oil route, has transformed the city into a commercial and administrative hub. Goats and cattle are exported from this port to Ethiopia and Yemen. The road linking Berbera to the capital, Hargeisa, is one instance of the area's infrastructure that requires maintenance. Despite attempts to raise funds through crowdsourcing, only a small portion of the necessary funds have been collected. Squatters are taking advantage of the many abandoned older Ottoman and merchant buildings.

Approximately 545,000 people currently reside in the Berbera metro area. Small-scale commercial ventures include selling ghat and frying fish. Many people rely on the diaspora who live abroad to send money home. Nevertheless, water sports have taken over as a draw for the stunning beaches. The Sea of Aden is warm all year round and has an abundance of fish, making shore diving possible and providing fishermen with a steady source of income. Given that the port of Berbera contributes 50% of the GDP of Somaliland, the government is investing in it to make it secure and appealing.

7.2.7. Urbanization Problems in Developing Countries

The rate of urbanization is rising quickly in developing nations. This has caused a scarcity of jobs in certain countries. People apply for benefits and programs funded by the government, which raises unemployment rates. In developing nations, "businesses and governments cannot produce enough jobs to meet the demand of a fast-growing population." Urbanization is linked to a number of issues, including high population density, poor infrastructure, a scarcity of affordable housing, pollution, *flooding*, the emergence of slums, crime, traffic, and poverty.

7.2.7.1. High Population Density

The high rate of migration from rural areas is the root cause of this high population density issue. Since there is a severe lack of housing due to rapid population growth, there is overcrowding, traffic jams, pollution, a shortage of housing (slum and squatter housing), high rents, poor urban living conditions, a lack of infrastructure services, unemployment, poverty, and a high crime rate. The high population density in developing nations is impacted by all of these.

Keyword

Public housing is a form of housing tenure in which the property is usually owned by a government authority, either central or local.

7.2.7.2. Inadequate Infrastructure

The state of the infrastructure is one of the major and serious aspects of the urban problem. Unreliable power supplies, poor road networks, insufficient water supplies, and a lack of basic amenities persist in certain developing nations.

7.2.7.3. Lack of Affordable Housing

Confrontations with well-organized squatters, who occupy vacant buildings to live rent-free or stop demolitions, have resulted from a lack of affordable housing. This has led to a low housing vacancy rate because of the quick increase in employment and rising housing costs. Issues like inadequate housing, which is particularly problematic for low-income families, are being dealt with. This has led to overcrowding in already crowded areas, the ongoing decay of dilapidated neighborhoods, significant social costs, and untold personal suffering. Less restrictive zoning laws, lower tax burdens, collaboration with private builders, encouragement of cooperative housing organizations, promotion of industrialized building techniques, use of inexpensive building materials, and more affordable mortgage credit are some of the strategies put forth to counteract the rising costs of *public housing*.

7.2.7.4. Flooding

Especially in developing nations, flooding is a major issue that urban areas must deal with during the rainy seasons. In certain developing nations, the inadequate construction of drainage systems causes problems when it comes to road accessibility during floods, ultimately resulting in disastrous flooding.

7.2.7.5. Creation of Slum

One of the major issues brought about by fast industrialization and urbanization is the expansion of slums in urban areas. In order to eradicate slums in urban societies, the following initiatives are proposed: a comprehensive slum clearance program; extensive rehabilitation and rebuilding; a slum improvement scheme; the building of an expressway network to mitigate urban decay and guarantee the central areas' continued prominence; the relocation of residents to the suburbs; and a slum upgrading scheme that provides some basic amenities, like street lights, drainage, and accessible roads.

7.2.7.6. Crime

The rate of crime rises in tandem with population growth. Most people living in urban cities suffer from high crime rates as a result of the high rates of unemployment, idleness, and joblessness in these areas. "Crime is inevitable in environments with poor living conditions and no economic security."

7.2.7.7. Congestion

It might be stuck in heavy traffic or crowded. Movement is challenging due to the route's overcrowding, which results in a slow and inefficient flow. Traffic congestion is a result of significant increases in the use of motor vehicles. Urban traffic congestion limits accessibility. Many laws, ordinances, and signage that are used to maintain orderly traffic flow can be used to regulate traffic flow. The way traffic moves helps to prevent accidents and other dangers.

Overcrowding, unemployment, housing problems, sanitary problems, water shortages, and health risks are some of the issues associated with urbanization. The process by which people move from rural to urban areas in quest of better employment prospects, healthcare, education, and a higher standard of living is known as urbanization. But it has led to socioeconomic problems that are now a big worry.

Did you get it?

1. **What are the main characteristics of urban settlement?**
2. **Give two examples of urban settlement.**

ROLE MODEL

Christopher Wren

English architect

Christopher Wren (born October 20, 1632, East Knoyle, Wiltshire, England—died February 25, 1723, London) was a designer, astronomer, geometrician, and the greatest English architect of his time. Wren designed 53 London churches, including St. Paul's Cathedral, as well as many secular buildings of note. He was a founder of the Royal Society (president 1680–82), and his scientific work was highly regarded by Isaac Newton and Blaise Pascal. He was knighted in 1673.

Early Academic Career and Scientific Pursuits

Wren was the only surviving son of a rector, and from an early age, he was delicate in health. Before Christopher was three, his father was appointed dean of Windsor, and the Wren family moved into the precincts of the court. It was among the intellectuals around King Charles I that the boy first developed his mathematical interests. The life at Windsor was rudely disturbed by the outbreak of the English Civil Wars in 1642. The deanery was pillaged, and the dean was forced to retire, first to Bristol and then to the country home of a son-in-law, William Holder, in Oxfordshire. Wren was sent to school at Westminster but spent much time under Holder's tuition, experimenting in astronomy. He translated William Oughtred's work on sundials into Latin and constructed various astronomical and meteorological devices. If the general direction of his studies was toward astronomy, however, there was an important turn toward physiology in 1647 when he met the anatomist Charles Scarburgh. Wren prepared experiments for Scarburgh and made models representing the working of the muscles. One factor that stands out clearly from these early years is Wren's disposition to approach scientific problems by visual means. His diagrams that have survived are beautifully drawn, and his models seem to have been no less elegant.

In 1649, Wren went to Wadham College, Oxford, as a "gentleman commoner," a status that carried certain privileges, and graduated with a B.A. in 1651. Oxford at that time had passed through a rigorous purgation of its more conservative elements by the parliamentary government. New men had been introduced, some of whom possessed great ability and had a special interest in the "experimental philosophy" so eloquently heralded by the scientific philosopher Sir Francis Bacon.

Receiving his M.A. in 1653, Wren was elected a fellow of All Souls College, Oxford, in the same year and began an active period of research and experiment, ending with his appointment as Gresham professor of astronomy in Gresham College, London, in 1657. In the following year, with the death of Oliver Cromwell and the ensuing political turmoil, the college was occupied by the military, and Wren returned to Oxford, where he probably remained during the events that led to the restoration of Charles II in 1660. He returned to Gresham College, where scholarly activity resumed and an intellectual circle proposed a society "for the promotion of Physico-Mathematicall Experimental Learning." After obtaining the patronage of the restored monarchy, this group became the Royal Society, with Wren being one of the most active participants and the author of the preamble to its charter.

In 1661, Wren was elected Savilian professor of astronomy at Oxford, and in 1669, he was appointed surveyor of works to Charles II. It appears, however, that, having tested himself successfully in so many directions, he still, at 30, had not found the one in which he could find complete satisfaction.

Turn to Architecture

One of the reasons why Wren turned to architecture may have been the almost complete absence of serious architectural endeavor in England at the time. The architect Inigo Jones had died about 10 years previously. There were perhaps half a dozen men in England with a reasonable grasp of architectural theory but none with the confidence to bring the art of building within the intellectual range of Royal Society thought—that is, to develop it as an art capable of beneficial scientific inquiry. Here, for Wren, was a whole field, which, given the opportunity, he could dominate—a field in which the intuition of the physicist and the art of a model maker would join to design works of formidable size and intricate construction.

An opportunity came, for in 1662 he was engaged in the design of the Sheldonian Theatre at Oxford. This, the gift of Bishop Gilbert Sheldon of London to his old university, was to be a theater in the classical sense, where university ceremonies would be performed. It followed a classical form, inspired by the ancient Theatre of Marcellus in Rome, but was roofed with timber trusses of novel design, thereby combining the classical point of view with the empirical modern in a way entirely characteristic of a Royal Society mind. At the same time, Sheldon probably was consulting Wren about London's battered—and in parts nearly derelict—St. Paul's Cathedral. So Wren was drawn, deeply and immediately, into building problems. What he desperately needed at that moment was contact with the European tradition of classicism, and he seized a chance to join an embassy proceeding to Paris.

By 1665 architecture at the court of Louis XIV had reached a climax of creativity. The Louvre Palace was approaching completion, and the remodeling of the Palace of Versailles had begun. Gian Lorenzo Bernini, the great sculptor and architect, was in

Paris making designs for the Louvre's east front, and the aged Italian allowed Wren to peruse his drawings. There was considerably more for Wren to see in the French capital, including the domed churches of the Val-de-Grâce and the Sorbonne and a marvelous array of châteaus within easy range of Paris.

By 1665 architecture at the court of Louis XIV had reached a climax of creativity. The Louvre Palace was approaching completion, and the remodeling of the Palace of Versailles had begun. Gian Lorenzo Bernini, the great sculptor and architect, was in Paris making designs for the Louvre's east front, and the aged Italian allowed Wren to peruse his drawings. There was considerably more for Wren to see in the French capital, including the domed churches of the Val-de-Grâce and the Sorbonne and a marvelous array of châteaus within easy range of Paris.

At Oxford in the spring of 1666, he made his first design for a dome for St. Paul's. It was accepted in principle on August 27, 1666. One week later, however, London was on fire. The Great Fire of London reduced two-thirds of the City to a smoking desert and old St. Paul's Cathedral to a ruin. Wren was most likely at Oxford at the time, but the news, so fantastically relevant to his own future, drew him at once to London. Between September 5 and 11, he ascertained the precise area of devastation, worked out a plan for rebuilding the City on new and more regular lines, and submitted it to Charles II. His plan reflected both his familiarity with Versailles and his acquaintance, through engravings, with the Rome of Pope Sixtus V. Others also submitted plans, and the king proclaimed on September 13 that a new plan for London would be adopted. No new plan, however, proceeded any further than the paper on which it was drawn. The problems of survey, compensation, and redistribution were too great. A rebuilding act was passed in 1667. It allowed only for the widening of certain streets, laid down standards of construction for new houses, levied a tax on coal coming into the Port of London, and provided for the rebuilding of a few essential buildings.

In 1669, the king's surveyor of works passed away, and Wren was promptly appointed. In December, he married Faith Coghill and relocated to the surveyor's official residence at Whitehall, where he resided, as far as is known, until his dismissal in 1718.

In 1670, a second rebuilding act was enacted, increasing the tax on coal and thereby providing a source of funds for the reconstruction of St. Paul's Cathedral and several churches within the City of London, as well as the construction of a column (The Monument) to commemorate the Great Fire. The city was now undergoing rapid reconstruction. Wren himself was not directly involved in the overall process. He occasionally offered advice to the City authorities on their major projects but did not design any houses or City companies' halls. He served as the king's surveyor based in Whitehall, not as an official of the City of London. St. Paul's and the City churches did not automatically fall under the royal works, although there was a longstanding tradition of royal responsibility for St. Paul's.

In 1670, the first churches were reconstructed. Eighty-seven churches had been destroyed in the fire, but some parishes were merged, resulting in only 52 being rebuilt.

While Wren was personally accountable for all of them, it should not be assumed that each one represents his fully developed design. The surviving drawings indicate significant delegation. Only a few are in Wren's handwriting. However, there is no doubt that Wren endorsed the design in each case, and in certain churches, his distinctive personality is evident.

Construction of St. Paul's

While the churches were being built, Wren was slowly and painstakingly developing designs for St. Paul's. The initial stage is represented by the First Model of 1670, now in the trophy room at the cathedral. This plan was approved by the king, and demolition of the old cathedral began. By 1673, however, the design seemed too modest, and Wren met his critics by producing a design of spectacular grandeur. A wooden model was made of this, and the Great Model, as it is called, is still preserved at St. Paul's. It failed to satisfy the canons of St. Paul's and clerical opinion generally, however, and Wren was compelled to withdraw from the ideal and compromise with the traditional. In 1675, he proposed the rather meager Classical-Gothic Warrant Design, which was at once accepted by the king, and within months building started.

What happened then is somewhat of a mystery. The cathedral that Wren started to build bears only a slight resemblance to the Warrant Design. A mature and superbly detailed structure began to rise. In 1694, the masonry of the choir was finished and the rest of the structure well in hand. In 1697, the first service was held in the cathedral. There was still, however, no dome. The building had been in progress for 22 years, and some restless elements in the government seemed to think this was too long. As an incentive for more rapid progress, half of Wren's salary was suspended until the cathedral was complete. Wren was now 65. Construction was completed in 1710, and in 1711, the cathedral was officially declared to be finished. Wren, at 79, petitioned for the withheld moiety of his salary, which was duly paid. The cathedral had been built in 35 years under one architect.

Concurrent Projects

Through all those years, Wren was not only the chief architect of St. Paul's and the City churches but also the head of the King's Works and thus the responsible officer for all expenditure on buildings issuing from the royal exchequer. He had an able staff to look after routine maintenance, but much business passed through his hands, including the control of building developments in and around Westminster. About 1674, the University of Cambridge considered building a Senate House for purposes similar to those for which the Sheldonian Theatre had been built. Wren made designs, but the project was abandoned. The master of Trinity College, who had promoted the scheme, was disappointed, but he persuaded his own college to undertake the erection of a new library (1676–84) and to employ Wren to design it. Wren's classicism here is impressive. There is no hint of the Baroque style prevalent in Europe at the time, and the building could well be mistaken for a Neoclassical work of a century later.

At Oxford in 1681, the dean of Christ Church invited Wren to complete the main gateway of the college. The lower part of Tom Tower, as the gateway was called, had been built by Thomas Cardinal Wolsey in a richly ornamental Gothic style. The octagonal tower that Wren imposed illustrates both his respect for Gothic and his reservations about it. His attitude toward Gothic design was consistent and influenced Gothic construction in England well into the 18th century. In 1682, Charles II founded the Royal Hospital at Chelsea for the reception of veterans superannuated from his standing army. The idea doubtless derived from Louis XIV's Hôtel des Invalides (1671–76) in Paris, but Wren's building, completed about 1690, is very different from its prototype. Charles II died in 1685. In the short reign of his brother, James II, Wren's attention was directed mainly to Whitehall. The new king, a Roman Catholic, required a new chapel; he also ordered a new privy gallery and council chamber and a riverside apartment for the queen. All these were built by Wren but were destroyed in the Whitehall fire of 1698.

There is not much information about Wren's personal life after 1669. He was knighted in the year of the Great Model, 1673. His first wife died of smallpox in 1675, leaving him with one young son, Christopher (another had died in infancy). His second wife was Jane Fitzwilliam (Fitz William), by whom he had a daughter, Jane, and a son, William, who died in 1679. In these years he never wholly abandoned his scientific pursuits. He was still at the center of the Royal Society and was its president from 1680 to 1682. He was sufficiently active in public affairs to be returned as a member of Parliament for Old Windsor in 1680 and, although he did not again take his seat, in 1689 and 1690.

With the Glorious Revolution of 1688, which drove James II from the throne, Wren found himself chief architect to William of Orange. William III and Mary II proved to be the most active builders of them all. They disliked Whitehall Palace, and in 1689 Wren was at work reconstructing two palaces: one at Kensington on the outskirts of London and the other at Hampton Court, 15 miles (24 km) away, up the River Thames. Kensington Palace was a piecemeal conversion of an older house, with new courts and galleries added. It is not a totally satisfactory composition, but the south front is a noble piece of brickwork. Hampton Court Palace, on the other hand, started as a project of huge dimensions—nothing less, in fact, than a rebuilding of the entire palace begun by Wolsey. Wren's first designs have survived, and in these he is seen, for the first time, spreading his wings as a palace architect. It was decided to demolish only half of the old palace, however, and Wren's design was reduced considerably. Nevertheless, he brought to it many innovations and a unique use of English building materials. Hampton Court is a mixture of red and brown brick and Portland stone combined in masterly equilibrium.

Queen Mary died in 1694. The king lost heart, and the building at Hampton Court was suspended; the palace was not completed until 1699. Two years before her death, the queen had initiated a scheme for the building of a royal hospital for seamen at Greenwich. For this, Wren made his first plans in 1694. The work began in 1696, but the whole group of buildings was not completed until several years after his death. Greenwich Hospital (later the Royal Naval College) was Wren's last great work and the only one still in progress after St. Paul's had been completed in 1710.

Queen Anne granted him a house at Hampton Court. He had, besides, a London house on St. James's Street, and it was there that a servant, noticing that he was taking an unusually long nap after dinner one evening, found him dead in his chair. Wren was buried with great ceremony in St. Paul's Cathedral, the tomb covered by a simply inscribed slab of black marble. On a nearby wall, his son later placed a dedication, including a sentence that was to become one of the most famous of all monumental inscriptions: "Lector, si monumentum requiris, circumspice" ("Reader, if you seek a monument, look about you").

Legacy of Christopher Wren

At his death, Wren was 90. He had far outlived the age to which his genius belonged. Even the men he had trained and who owed much to his original and inspiring leadership were no longer young. The Baroque school they had created was already under fire from a new generation that brushed Wren's reputation aside and looked back beyond him to Inigo Jones. Architects of the 18th century could not forget Wren, but they could not forgive those elements in his work that seemed to them unclassical. The churches left the strongest mark on subsequent architecture. In France, where English architecture rarely made much impression, St. Paul's Cathedral could not be easily ignored, and the Church of Sainte-Geneviève (now the Panthéon) in Paris, begun about 1757, rises to a drum and dome similar to St. Paul's. Nobody with a dome to build could ignore Wren's, and there are myriad versions of it, from St. Isaac's Cathedral (dome constructed 1840–42; completed 1858) in St. Petersburg to the U.S. Capitol at Washington, D.C. (dome built 1855–63).

It was only in the 20th century that Wren's work ceased to be a potent and sometimes controversial factor in English architectural design. The last major architect to have been confessedly dependent on him was Sir Edwin Lutyens, who died in 1944. The Wren Society, founded at the bicentenary of Wren's death in 1923, published 20 volumes of Wren material (1924–43), edited by A.T. Bolton and H.D. Hendry.

FOCUS ON CAREERS

URBAN PLANNER

A specialist in town and city planning and designing is known as an urban planner. They are responsible for neighborhoods, public spaces, industrial zones, parks, streets, and other areas. Urban planners determine the most effective way to use the available land while taking into account the needs of city dwellers. They decide how to rebuild and improve existing neighborhoods to improve daily activities by speaking with residents about their needs and gathering information. To revitalize infrastructure and accommodate a growing population, they also need to consider potential future urban development.

Role of An Urban Planner

Aside from the primary task of developing urban projects on available land, the work duties of an urban planner include the following:

- Analysing censuses, market data and environmental and economic studies
- Assessing the size and potential use of available land for development
- Investigating and analyzing various factors that could affect land use
- Meeting the public, government officials and developers regarding land development
- Visualizing and planning roads, buildings, parks and other urban features
- Reviewing and approving site plans for the land under consideration
- Submitting urban projects for approval to urban planning officials and commissions
- Adhering to local, state and national laws regarding urban development

What Do You Need To Become An Urban Planner?

A variety of hard and soft skills are required to work as an urban planner. Strong verbal and written communication skills are crucial for urban planners because they work in teams and frequently engage with the public, developers, and government officials. Excellent public speaking, presentation, and diplomatic skills could be a career advantage. You need to be technologically savvy to use spreadsheets, database applications, geographic information systems (GIS), and other pertinent software to complete your daily work tasks. Additionally, gathering, organizing, and analyzing statistical data are critical skills.

Skills of an Urban Planner

Urban planners generally require the following skills to thrive in their profession:

- Engineering and mathematical skills: Planning, designing, constructing, and renovating streets, buildings, and land require a thorough grasp of engineering and mathematical concepts.
- Planning abilities: To turn an area into a useful area for urban residents, urban planners must evaluate the area's potential and develop practical plans.
- Creativity: An urban planner with a creative mindset can visualize urban environments and structures.
- Computer skills: To plan, design, and carry out a variety of urban projects, urban planners must be proficient with computers and computer programs.
- Communication: An urban project's success depends on its participants' ability to understand, discuss, and convey plans and instructions to one another.
- Decision-making abilities: As part of their job, urban planners assess all of the options available and choose the option that best fits the requirements, usage, safety regulations, and budget.
- Leadership: An urban planner needs to possess strong leadership abilities in order to manage large projects, guide the participants in them, and achieve project objectives.
- Interpersonal skills: Urban planners can gain from having strong interpersonal skills since they must communicate and collaborate with government representatives, developers, project workers, and other individuals.

Majoring in geography, urban design, regional planning, or urban planning are options for urban planners. If you have a degree in architecture, transportation engineering, environmental engineering, or civil engineering, you can also work in this field. Candidates with a master's or bachelor's degree are typically preferred by employers.

How To Become An Urban Planner

You can become an urban planner by taking the following steps:

Complete 10+2

It is essential to pass the 10+2 in the science stream, with physics, chemistry and mathematics. To be eligible for higher studies in urban planning, a base in mathematics is compulsory. Try to get more than the minimally required 55% aggregate marks.

Pass Competitive Exams

After earning the Class 12th Higher Secondary Certification (HSC), you can appear for competitive college entrance exams. These include the National Aptitude Test in

Architecture (NATA) and the Joint Entrance Examination–Main (JEE Main). Passing these exams can make you eligible for various college courses.

Get a Bachelor's Degree

You can get a Bachelor of Planning (B.Planning) degree or a Bachelor of Architecture (B.Arch) degree. Other graduate degree programs suitable for an urban planner career include economics, geography, environmental design, landscape architecture, civil engineering, public administration, and political science. With these four-year degrees, you can get an entry-level urban planning position in various public and private sectors.

Earn a Master's Degree

Higher educational credentials may help you to get better career opportunities. You can benefit from a Master of Planning (M.Plan) in Urban Planning or a Master of Planning (M.Plan) in Urban and Regional Planning. You are eligible for these two-year programs if you have a relevant graduate degree and a passing score on the Graduate Aptitude Test in Engineering (GATE) exam.

Apply for Jobs

You can apply for urban planning jobs with municipal corporations, state and national government departments, construction companies, and land consultancy firms. Passing civil services exams like the Union Public Service Commission (UPSC) may help you get a government position. It is advisable to prepare a resume and cover letter and practice answering commonly asked interview questions.

Salary

The national average salary of urban planners is $38,540 per year. The exact pay may vary according to educational qualifications, skills, experience, employer and location. The pay scale may go up as you gain more professional experience.

Work Environment

In addition to visiting various project locations regularly, an urban planner may work in an office. In addition to meeting with developers, architects, engineers, and public officials, they carry out planning, designing, and presenting tasks in the office. They conduct research and make assessments while on the scene. An urban planner can verify that their project plans are realistic and accurate by working in the field. Typically, urban planners are employed by state and federal government agencies, local governments, building firms, consulting firms, non-profit organizations, and real estate developers.

A CLOSER LOOK

UNEMPLOYMENT

Being unemployed means having the ability to work and actively looking for work, but not being able to find any. It's crucial to remember that a person must be actively seeking paid employment in the labor force in order to be classified as unemployed.

The term "underemployment" refers to the circumstances of individuals who can only secure work for shorter than usual periods, such as part-timers, seasonal workers, or temporary or day workers. The term can also apply to employees who are overqualified for their positions due to their education or training.

In most countries, government labor offices collect and analyze data on unemployment, which is now widely considered a key indicator of the state of the economy. Unemployment trends and demographic statistics are scrutinized to understand broader economic patterns and to guide potential policy measures. Since World War II, achieving full employment has been a stated goal of many governments, leading to the development of various initiatives. It is important to note that full employment does not always mean a zero unemployment rate, as there will always be individuals in the labor force who are in transition between jobs but not officially classified as unemployed.

SUMMARY

- Rural communities are closely tied to the land, with primary industries such as farming, fishing, and animal husbandry being dominant. These settlements are typically small in size and can be categorized into four types: hamleted, semi-compact, compact, and dispersed or scattered.
- A building's or house's relationship to another is known as the settlement pattern, which can be identified by examining a map at the local scale. The pattern covers only compact and semi-compact settlements; dispersed settlements have a distinct form.
- Rural settlements in under-developed countries lack infrastructure and are numerous. An urban settlement is a densely populated area where people live and work in urbanized surroundings.
- A regional plan usually divides cities into regulated zones according to requirements for particular industries, such as environment, habitat, housing, and sanitation.
- Typical classification factors of urban settlement include aspects like administrative structure, occupation structure, and population size.
- Urban areas are classified as cities based on their size, services offered, and functions performed.
- A place's historical and contemporary ties to other locations impact its social and economic features.
- The rate of urbanization is increasing rapidly in developing nations. This has led to a shortage of jobs in some countries. As a result, people are seeking benefits and programs funded by the government, which in turn contributes to higher unemployment rates.

MULTIPLE CHOICE QUESTIONS

1. **Urbanization is best defined as:**
   a. People moving from rural areas to urban areas
   b. The growth in the population of urban areas as a result of several factors
   c. The increase in the proportion of people living in urban areas
   d. People moving from urban areas to rural areas
2. **Throughout most of history:**
   a. Half the world's population lived in urban areas
   b. Most of the world's population lived in urban areas
   c. 51% of the world's population lived in urban areas
   d. Most of the world's population lived in rural areas

3. **All of the following are reasons for urbanization EXCEPT:**
   a. Rent is more affordable in urban areas
   b. Urban areas have better healthcare and educational facilities
   c. There are more jobs available in urban areas
   d. There are more opportunities for entertainment in urban areas
4. **Which of the following statements about urban areas is NOT true?**
   a. There is traffic congestion
   b. Public utilities are not readily available
   c. There are higher crime rates
   d. Slums often develop in and around urban areas
5. **Cities often spread out over the surrounding areas. The is called_________.**
   a. Urban spread
   b. Urban growth
   c. Urban overflow
   d. Urban sprawl
6. **The movement of people from one country to another is called__________.**
   a. Migration
   b. Rural to urban migration
   c. International migration
   d. Annual migration
7. **A person moves to another country to study for a few years and then returns home. This is an example of:**
   a. Short term migration
   b. Long term migration
   c. Rural to urban migration
   d. Annual migration
8. **For most Caribbean territories:**
   a. The number of immigrants is greater than the number of emigrants
   b. The number of immigrants is roughly equal to the number of emigrants
   c. The number of immigrants is much less than the number of emigrants
   d. The number of emigrants is insignificant
9. **Inadequate educational and health infrastructure for; large rural populations is a big issue in under-developed countries.**
   a. True
   b. False

10. Urbanization is linked to a number of issues, including high population density, poor infrastructure, a scarcity of affordable housing, pollution, flooding, the emergence of slums, crime, traffic, and poverty.

a. True

b. False

REVIEW QUESTIONS

1. What is human settlement? Explain in detail.
2. What are the factors influencing rural settlement?
3. What are the types of urban settlements?
4. What is the big segment for defining urban and rural areas?
5. Give an example of polycentric city.
6. What are the main causes of urbanization?

Answer to Multiple Choice Questions

1. (c)	2. (d)	3. (a)	4. (b)	5. (d)
6. (c)	7. (a)	8. (c)	9. (a)	10. (a)

REFERENCES

1. Allen, A., & You, N., (2002). *Sustainable Urbanization: Bridging the Green and Brown Agendas*. Development Planning Unit University College London. London.
2. Brebbia, C., Martin-Duque, J., & Wadhwa, L., (2002). *The Sustainable City II. Urban Regeneration and Sustainability*. WIT Press. Boston.
3. Buckley, R., & Mini, F., (2000). *From Commissars to Mayors – Cities in the Transition Economies*. Washington, D.C. World Bank.
4. Correa, C., (2000). *Housing and Urbanization*. London: Thames & Hudson.
5. Crowe, M., Nolan, K., Collins, C., & Kristoffersen, M., (2002). *Biodegradable Municipal Waste Management in Europe: Strategies and Instruments.* European Topic Centre on Waste.
6. Fiedler, J., (2014). *Urbanization, Unlimited*. Cham: Springer International Publishing. http://dx.doi.org/10.1007/978-3-319-03587-1.
7. Katz, B., & Lang, R., (2003). *Redefining Urban and Suburban America: Evidence from Census 2000* (Vol. 1). Brookings Institutions Press. Washington, D.C.
8. Witherick, M., (2006). *Cities & Urbanization*. Deddington: Philip Allan Updates.

CHAPTER 8

Human Development

LEARNING OBJECTIVES

After studying this chapter, you will be able to:

- Understand growth v/s development
- Identify approaches to human development
- Discuss about human development index
- Explain the dimensions of human development
- Describe the role of human capital in economic development

"Economic growth and human development need to go hand in hand. Human values need to be advocated vigorously."

—Kailash Satyarthi

INTRODUCTORY EXAMPLE

Economic Development

Economic development is the process through which low-income, simple national economies are changed into sophisticated, industrialized economies. The phrase is typically used to characterize a shift in a nation's economy encompassing both qualitative and quantitative advancements, though it is occasionally used as a synonym for economic growth. For underdeveloped nations, the theory of economic development—how rudimentary and impoverished economies can transform into sophisticated and relatively prosperous ones—is crucial, and this is typically the context in which the topic of economic development is addressed.

Post-World War II, economic development emerged as a significant concern. As European colonialism came to an end, many former colonies and other low-living nations were referred to as "underdeveloped countries" to distinguish their economies from those of the developed nations, which were perceived to be those of the US, Canada, the UK, western and eastern Europe, the former Soviet Union, Japan, South Africa, Australia, and New Zealand. The impoverished countries were dubbed developing countries as their living standards improved over the next few decades.

The definition of a developing country and the elements that make up the process of economic development are not agreed upon by all. Per capita income is typically used to classify developing nations, and it is generally accepted that economic development happens as per capita income increases. The most accurate indicator of the value of the goods and services that the society provides to each individual member annually is its per capita income, which is almost equal to its per capita output. These two indicators are the best available for providing estimates of a country's economic growth and level of well-being, although measuring per capita income and its rate of growth both have some issues.

It is advisable to take into account certain statistical and conceptual challenges associated with applying the traditional underdevelopment criterion before delving into the root causes of underdevelopment. Statistics present well-known challenges. First, there are the awkward cases that fall in between. Even if the analysis is limited to developing and underdeveloped nations in Latin America, Africa, and Asia, there are wealthy oil-producing nations that have per capita incomes significantly higher than the average but lack developed general economic traits. Second, though expressed in terms of an international currency like the U.S. dollar, a very imprecise indicator of their actual per capita income, many developing countries' per capita incomes are rendered unreliable due to a variety of technical issues. Among these challenges are the inadequacies in the fundamental national income and population statistics, as well as the inappropriateness of the official exchange rates used to convert national incomes in terms of the various domestic currencies into the common denominator of the U.S. dollar, as well as the challenges associated with determining the worth of the non-cash parts of real incomes in developing nations.

Measures of per capita income have a reasonable correlation with other indicators of economic well-being, such as life expectancy, infant mortality rates, and literacy rates, despite the well-established challenges with income measures. There is a strong correlation between per capita income levels and other indicators, including the state of nutrition and the availability of doctors, teachers, and hospital beds per capita. The actual observed differences are much larger in magnitude, even though a difference of, say, 10% in per capita income between two countries would not be considered as necessarily indicative of a difference in living standards between them. For instance, in 1985, the per capita income of India was estimated to be $270. It was estimated to be $6,520 for Italy, while it was $1,640 for Brazil. No one would dispute that the living standard in Italy was significantly higher than that of Brazil, which was higher than India's by a significant margin, even though economists have pointed out a number of reasons why the implication that Italy's living standard was 24 times greater than India's might be biased upward.

There are two caveats to accepting the idea that a low per capita income level is a material indicator of poverty. First, the standard of living is determined by per capita consumption rather than per capita income. If a significant amount of the national income is diverted from consumption for other reasons—such as a policy of forced saving—the two may diverge significantly. Second, the representative standard of living of a nation's vast majority of citizens more accurately represents the level of poverty within that nation. When there is a significant disparity in the standard of living between the rich and the poor and the national income is highly unequally distributed, this may be significantly lower than the simple arithmetic average of per capita income or consumption. The World Bank's definition of a developing nation is generally accepted: "low-income developing countries" were those that had per capita incomes under $400 in 1985, and "middle-income developing countries" were those that had per capita incomes between $400 and $4,000. Even so, nations with similar per capita incomes might not be alike in other ways. For example, some nations might get a large portion of their income from capital-intensive industries like oil extraction, while other nations with comparable per capita incomes might employ their labor force in more creative and fruitful ways to make up for their lack of abundant natural resources. For instance, Kuwait's per capita income was estimated to be $14,480 in 1985, but half of that amount came from oil. Kuwait's social and economic indicators were generally far lower than those of other nations with comparable per capita incomes. Although China and North Korea are universally regarded as developing countries, centrally planned economies are also generally regarded as a distinct class. One significant challenge is that prices are not as good a gauge of relative scarcity in the central domain.

There is a marginally smaller margin of error for estimates of percentage increases in real per capita income than there is for estimates of income levels. Growth rates of per capita income over periods of a decade or more are strongly indicative of the rate at which average economic well-being has increased in a country, even though year-to-year changes in per capita income are heavily influenced by factors like weather (which affects agricultural output, a large component of income in most developing countries), a country's terms of trade, and other factors.

INTRODUCTION

The process of increasing people's chances and freedoms while also enhancing their well-being is known as human development. Human development is the actual freedom that common people have to choose their identities, careers, and lifestyles.

Capabilities are a key concept in the human development approach. The tools needed to pursue a meaningful life are capabilities, or what people can do and become. Almost everyone values three basic capabilities: having access to knowledge, being in good health, and having a respectable standard of living. Other essential qualities for a happy life could be being able to take part in decisions that impact one's life, control one's living environment, be free from violence, be respected by society, and be able to unwind and enjoy yourself.

Both the structures and norms of our society and our own efforts can increase (or limit) our potential. The means to turn their idea of "a good life" into a reality are available to those with broad, highly developed capabilities. People with low ability are less able to take advantage of opportunities and forge their own path. Human potential cannot be realized in the absence of fundamental abilities.

One normative framework for analyzing well-being that is frequently used to comprehend development issues is the capability approach. The development and spread of the approach can be traced back to philosopher-economist Amartya Sen and, more recently, Martha Nussbaum, a professor of law and ethics at the University of Chicago, even though some aspects of it can be traced back to Aristotle and Adam Smith. The fundamental tenet of the capability approach is that people's capabilities—often referred to as their "capabilities to function"—should be defined by their actual and genuine opportunities to pursue their desired endeavors and, via these freedoms, to be whoever they choose to be. The concept of educational opportunity serves as one example of the distinction between formal freedoms and functional capacities. Formally, every citizen of the United States is entitled to a college degree. However, the actual freedom of low-income students to attend college can be restricted by a variety of factors, including low-quality local high schools and financial concerns, when comparing them to more affluent students from similar neighborhoods. Formal liberties are necessary but insufficient in this and many other situations.

8.1. GROWTH V/S DEVELOPMENT

Learning Objectives

- Understand growth and development differences
- Explain growth and development characteristics
- Give growth and development examples
- Identify factors affecting growth and development
- Discuss barriers to growth and development
- Know the policies for growth and development

Development and economic growth are both critical to the long-term viability of a nation's economy. Economic growth is typically positive because it indicates that the economy is producing more goods and services, which raises output.

8.1.1. Growth and Development Differences

There are a few significant distinctions between the concepts of economic development and growth, despite their apparent similarities. An increase in the overall output of a nation's economy is known as economic growth. It has to do with how much is produced in the economy over time in terms of goods and services.

A nation's economy is typically doing well when it is growing economically. It does not, however, always translate into an improvement in the nation's overall standard of living or **economic prosperity**. Here's where economic development can serve as a useful metric.

Economic development evaluates the economy's changes in output as well as the quality of that output.

Therefore, we can assess economic welfare and general happiness in the economy using economic development. Qualitative aspects are also taken into consideration in economic development; these include things like rising living standards and easier access to healthcare and education, to mention a few (Figure 8.1).

Keyword

Economic prosperity refers to a country's economic growth, security, and competitiveness.

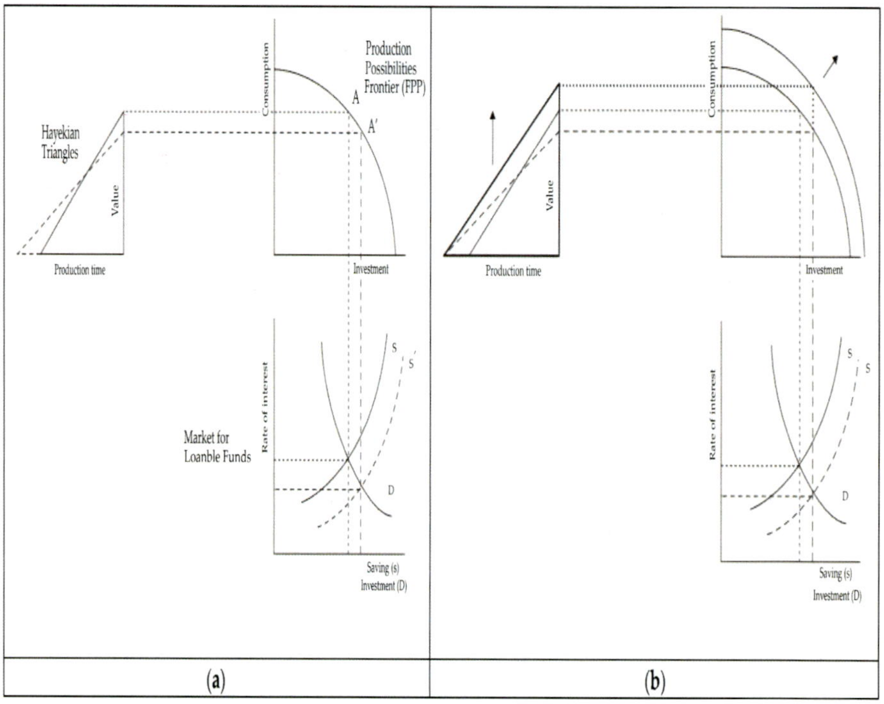

Figure 8.1. The dynamic efficiency of sustainable economic growth and development.

Source: Espinosa VI, Alonso Neira MA, Huerta de Soto J. Principles of Sustainable Economic Growth and Development: A Call to Action in a Post-COVID-19 World. *Sustainability*. 2021; 13(23):13126. https://doi.org/10.3390/su132313126.

8.1.2. Growth and Development Characteristics

The Rostow model helps us comprehend the features of growth and development. Throughout the 1960s, W. W. Rostow created a model that contends developing nations experience five different phases. These five stages, each of which represents a feature of developing economies, are represented by a linear model.

"Traditional society" is the name of the initial phase. This phrase was coined by Rostow to characterize societies with stagnant technological advancements. He maintained that societies in this stage are essentially archaic, impoverished, and undergo little to no generational change. Their economic development is nonexistent. "Preconditions for take-off" is the next phase. Traditionally based economies are preparing for "take-off," or that is, they are starting to exhibit indications of industrial growth. To make room for trade, savings, and investment, this also entails the exploitation of agriculture and other primary-sector businesses.

"Take-off" refers to the third stage, which Rostow defines as self-sustaining growth. The establishment of a manufacturing sector and the construction of physical infrastructure are included in this stage. By the time this stage is over, contemporary political, social, and economic structures are in place, but economic development is viewed as secondary to economic growth. The "drive to maturity," which denotes the growth of a broad spectrum of industrial and commercial activities, is the fourth stage. Internationally, nations can now take advantage of their comparative advantages. When developing economies reach the last stage, known as "high mass consumption," they can finally fully benefit from self-sustaining growth. They can distribute the extra economic output toward social welfare, or more specifically, toward economic growth.

Remember

Economic development assesses not only the change in the output of the economy but also the quality of the output.

It is crucial to remember that Rostow's theory is fairly basic and was created a long time ago. Though it shouldn't be interpreted as gospel, it is a useful place to start when learning about the traits and phases of growth and development.

8.1.3. Growth and Development Examples

Now, let's look at some instances of growth and development while also examining the indicators and contributing factors.

Economic growth and development are frequently measured using the gross domestic product (GDP) and *gross national income* (GNI) per capita. This is due to the fact that per capita income rather than GDP and GNI figures combined offer a more accurate picture of the country's income. Since these statistics are based on economic output rather than welfare, they frequently still fail to provide a true picture of economic development. Understanding the three primary facets of economic welfare is crucial for determining more precise metrics of economic development. They are:

1. Welfare that has been acquired from the goods and services produced and purchased in the economy.
2. Welfare gained from public and merit goods.
3. Welfare derived from 'quality of life' factors.

Since they account for the entirety of economic welfare in a nation's national economy, these three factors taken together can be used as an indicator of welfare. Notably, sustainability is not taken into account by national income metrics. The degradation and exhaustion of natural resources brought on by the production of diverse goods and services are not factored into GDP or GNI calculations. Therefore, while a nation's economy may be expanding, its natural resources may be running out.

Keyword

The **gross national income**, previously known as gross national product, is the total domestic and foreign output claimed by residents of a country, consisting of gross domestic product, plus factor incomes earned by foreign residents, minus income earned in the domestic economy by nonresidents.

Measuring economic development is further complicated by the fact that neither GDP nor GNI offer reliable estimates of the output of merit goods. Because it considers health and education in addition to national income, the Human Development Index (HDI) is therefore frequently used as a gauge of economic development. The three components of the Human Development Index (HDI), a macroeconomic metric put forth by the UN, are GNI per capita, mean and expected years of education, and life expectancy. An HDI can have a maximum value of 1. For instance, the United Nations reports that the United Kingdom's HDI is 0.932, which is very near to 1.

Question

What is economic growth?

Answer

Economic growth consists of various processes in per-capita income and national income increasing process.

8.1.4. Factors Affecting Growth and Development

Three main factors impact growth and development:

- Investment: Long-term economic development—an improvement in the caliber of goods and services produced as a result of improved technology—and growth—an increase in the output of goods and services—depend on investments in capital goods and technology.
- Education: Development and economic growth both depend on this. People are more likely to find employment or even start their own businesses when they have high levels of education and training, which boosts the output of the national economy as a whole. Education also helps people become more intelligent. As a result, spending on education will boost economic growth as well as long-term economic well-being.
- Healthcare: Raising healthcare spending raises living standards. For example, one of the primary metrics utilized in the construction of the HDI is *life expectancy*. Investing in healthcare can increase life expectancy, which will raise general societal welfare.

8.1.5. Barriers to Growth and Development

Growth and development, unfortunately, are not always simple processes. An economy may encounter obstacles along the way that prevent it from expanding and developing sustainably.

- Corruption: Any dishonest or fraudulent behavior frequently displayed by those in positions of authority is considered corruption. Bribing someone is a typical instance of corruption. This results in an inefficient use of resources and obstructs regular economic activity, which hinders sufficient economic expansion and advancement.
- Institutional factors: Developing nations may occasionally lack some institutional elements found in developed nations. For example, less developed legal, judicial, or financial systems may make it difficult for businesses to fairly provide goods and services to consumers, which would impede economic growth.
- Infrastructure: Developing nations frequently lack the infrastructure required for economic expansion. For example, businesses may find it more difficult to obtain raw materials or to transport their finished goods if there are no roads or other long-distance transportation options. This could result in increased costs and slower development.
- Human and physical capital: Insufficient resources in these areas can also impede the expansion and advancement of the economy. As previously mentioned, investments in human capital, education, and training are all essential for economic expansion.

Keyword

Life expectancy is an estimate of the average number of additional years that a person of a given age can expect to live.

8.1.6. Policies for Growth and Development

To promote economic growth and development, a variety of macroeconomic and microeconomic policies may be used. Supply-side policies, monetary policies, and fiscal policies are a few types of potential policies.

Finding the right policy to put into place is not an easy task, though. Ensuring the sustainability of economic growth is a crucial aspect to consider. Additionally, consideration should be given to the environment and the finite supply of natural resources. Resource depletion could eventually result in much more serious issues that outweigh the advantages of rapid economic expansion.. Therefore, policies that lessen the likelihood of market failure brought on by economic growth should be taken into account. Fiscal policies that promote the redistribution of income and wealth (i.e., the use of taxes to remove the negative externalities resulting from production) are examples of such policies, like *progressive taxation*. Most people consider the fair distribution of income to be a sign of economic development.

Pursuing economic development and growth could be challenging. In order to promote sustainable growth and development, effective policies must be carefully chosen because of the many variables that affect these processes, as well as the multiple trade-offs and obstacles that may arise when pursuing them.

Keyword

Taxation is a term for when a taxing authority, usually a government, levies or imposes a financial obligation on its citizens or residents.

Did you get it?

1. **What is the main difference between growth and development?**
2. **What are some important factors that determine economic growth and development?**

8.2. APPROACHES TO HUMAN DEVELOPMENT

Learning Objectives

- Discuss about equity
- Explain efficiency and productivity
- Focus on participation and empowerment
- Understand sustainability
- Know the capability approach to human development
- Identify levels of human development approaches

The approach to human development is multifaceted by nature. Enabling individuals to play an active role in their own lives is the primary goal of human development. Individuals are active subjects with the freedom to choose how they wish to live, not simply passive recipients of social welfare programs. They should have the autonomy to set their own priorities and determine the most effective ways to achieve them. Therefore, enhancing freedom and nurturing agency are interconnected. Individuals must have the freedom to learn, to speak publicly without fear, and to associate and express themselves freely to become agents of their own lives.

The four main pillars of human development are:

- equity;
- efficiency and productivity;
- participation and empowerment;
- sustainability.

8.2.1. Equity

The ideal of equality, according to which every person should have an equal voice in civic and political life as well as equal rights and entitlements to human, social, economic, and cultural development, is encompassed by the principle of equity. It also acknowledges the possibility that affirmative action, or preferential treatment, is necessary for people who have uneven opportunities as a result of different disadvantages. For instance, different people will experience different levels of utility from the same income or investment based on their individual characteristics, starting endowments, or conversion factors, which help convert inputs into outputs. The aim of equality is to guarantee that those who are denied access to fundamental opportunities like health and education do so because their opportunities differ from those of other segments of society. Therefore, equity seeks to achieve equality of resources—not just financial ones—but also of health, education, and *employment* opportunities, as well as democratic participation. Equitable outcomes follow from the achievement of the equal opportunities goal.

Keyword

Employment is a relationship between two parties regulating the provision of paid labor services.

8.2.2. Efficiency and Productivity

Efficiency is the least expensive way to achieve objectives by making the best possible use of institutional, material, and human resources to maximize opportunities for both individuals and communities and boost productivity. Building infrastructure such as roads and dams is one example of how effectively using limited national resources can benefit people's lives. Improving productivity can be achieved by using resources more effectively. To the fullest extent possible, people must also be invested in, and an enabling macroeconomic environment must exist. People must be given the tools they need to become more productive and fully engaged in the process of seizing opportunities for them to be effective growth agents and for human development to occur.

8.2.3. Participation and Empowerment

Engagement and participation in social and political life are crucial components of human development. Participation from the public is essential for government initiatives and community programs. People will participate more in decision-making when grassroots support is mobilized through decentralization in planning, as this approach brings the government closer to the people. Involvement can also help to improve the standard of social service delivery by allowing people to ask questions of authorities. It puts pressure on local authorities to address deficiencies or gaps in institutional performance as soon as they are discovered. Increased engagement and participation can lead to empowerment. For example, women's reservations in elected bodies are intended to empower them through their involvement. Another example is the participation of communities, parents, and/or guardians in village education committees.

8.2.4. Sustainability

Human development challenges the notion that economic growth can be sustained over the long run and works to ensure that resources are used in a way that satisfies current needs and protects the environment so that future generations' needs can also be met. Therefore, it is crucial to utilize resources without harming the environment to ensure that the progress made is enduring and has the potential for further growth. For instance, the development process cannot be sustainable in the long term if it fails to establish institutions that uphold people's rights.

8.2.5. Capability Approach to Human Development

Amartya Sen's groundbreaking work in welfare economics, social choice, poverty, famine, and development economics has had a significant impact on the human development approach. Sen's writings cover a wide range of subjects, but his "capability approach" has greatly influenced the study of economics and the social sciences overall. The capability approach finds its roots in the works of Karl Marx, Adam Smith, and Aristotle. Aristotle linked his exploration of human functions with his own analysis, drawing heavily from it. Karl Marx and Adam Smith discussed the importance of "functionings" and "capability" as measures of well-being. If life consists of meaningful actions and states of being, then assessing these functionings and functional capabilities is how one gauges the quality of life.

Keyword

A **capability** is something a person or a thing is able to do.

However, what precisely are the terms "functioning" and "capability"? Amartya Sen states that "functioning is a vector of functionings, reflecting the person's freedom to lead one type of life or another... .to choose from possible livings.." Put differently, capabilities refer to the actual freedoms that an individual has to live the kind of life that they find meaningful. A person with many abilities can enjoy a wide range of activities and follow a variety of life paths, just as someone with a large pocketbook can purchase a wide range of goods. A person's functionings are the worthwhile pursuits and conditions that make up their overall state of health, safety, education, and other aspects of life. Hence, functioning is a person's accomplishment—whatever it is that they can accomplish or become. When someone's basic need for food is satisfied, for instance, they can enjoy the benefits of being well-nourished. The third fundamental idea of the capability approach is "agency," which goes along with *capability* and functioning. It speaks to the capacity of an individual to seek and accomplish objectives that they have good reason to value.

Martha Nussbaum contends, however, that Sen's "Capability Approach" is unfinished. She contends that to ensure that everyone has equal freedom, one must get past these obstacles and move past this ambiguity. This is because what people value and find relevant can frequently be the result of systems of discrimination and inequality, and because not all human freedoms are created equal—for instance, the freedom to pollute is not the same as the freedom to protect the environment. Within this framework, she has put forth a list of ten essential human capacities that make up the public policy evaluation domain. The capability approach promotes removing barriers from people's lives to give them more flexibility to accomplish the functioning that they find important. It suggests progressive social policies that would guarantee fair access to human opportunities while also promoting the advancement of human capabilities, such as better health, knowledge, and skills.

Keyword

The human development approach conceives development as being about expanding people's well-being.

8.2.6. Levels of Human Development Approaches

Human development is divided into three stages. Three categories of development groups are established for each country: higher, medium, and lower-level development groups.

High-level Development Groups:

- Countries that rank high in the Human Development Index belong to this development group.
- These nations have higher rankings because they have made investments in important fields like healthcare and education. The governments of these nations also place a high premium on building the infrastructure of the social sector.
- There is a notably higher investment in providing good governance to people.
- Examples of such countries are Norway, Ireland, Iceland, Switzerland, Germany, and Sweden.

Medium Development Groups:

- There are at least 40 countries at the medium level of human development.
- Most of these countries were formed after the Second World War. Many of these countries were either colonized or were formed after the Soviet Union dissolved.
- These countries rank higher in the HDI when it comes to social diversity.

- Many of the countries in this development group have struggled with socio-political and economic instability.

Examples of medium-level development groups include Brazil, Mexico, India, Tunisia, Vietnam, and Morocco.

Lower Development Groups:

Forty countries are part of the lower-level development group. These are smaller countries that are experiencing socio-political instability. Many of these countries are either in the midst of civil war, famine, or drought.

Examples of lower-level development groups are Sudan, Liberia, Afghanistan, Syria, and Pakistan.

Development that focuses on people and enhancing their opportunities is known as human development. The income approach, welfare approach, capability approach, and basic needs approach are the four different approaches to human development. Additionally, the Human Poverty Index and the Human Development Index are the two indices most commonly used to measure human development.

Question

What is the capability approach?

Answer

The capability approach to human development is associated with acclaimed Indian economist Amartya Sen. Under the capability approach, the key to human development is building human capabilities in education, health, and accessibility of resources.

Did you get it?

1. **What is the approach to measuring human development?**
2. **What is the impact of HDI on a country?**

8.3. HUMAN DEVELOPMENT INDEX

Learning Objectives

- Understand the Human Development Index (HDI)
- Measure HDI
- Know HDI Rankings
- Discuss about limitations of the HDI

The Human Development Index (HDI) gives an overview of average performance in three key areas of human development: living a long and healthy life, having knowledge, and enjoying a reasonable standard of living. The geometric mean of the normalized indices for each of these dimensions is referred to as the HDI. The education dimension looks at the average years of schooling for adults aged 25 and over and the expected years of schooling for children about to start school. The health dimension is based on life expectancy at birth, while gross national income per capita acts as a measure for the standard of living. The HDI uses the logarithm of income to show how income's importance decreases as GNI increases. By combining the scores for the three HDI dimension indices using the geometric mean, a composite index is created.

The HDI can question national policy decisions by highlighting how two nations with similar GNI per capita can have very different human development outcomes. These differences can lead to discussions on public policy objectives. However, the HDI only captures a portion of what human development entails, simplifying complex issues like poverty, human security, inequality, and empowerment. To address major concerns like poverty, inequality, gender inequality, and human development more comprehensively, the HDRO provides additional composite indices. Analyzing extra indicators and data from the HDR statistical annex is crucial for a more complete understanding of a nation's human development status.

8.3.1. Understanding the Human Development Index (HDI)

The Human Development Index (HDI) was designed to prioritize individuals—or, more accurately, their potential to achieve fulfilling careers and personal lives. Beyond taking into account conventional measures of economic growth, like GDP, assessing a nation's potential for individual human development offers an additional metric for assessing its degree of development.

The HDI can be used to assess why two countries with roughly the same GNI per capita produce such drastically different results for human development. It can also be used to look at the different policy decisions made by different countries. The proponents of HDI believe this kind of fruitful public policy discourse can be fostered by the tool (Figure 8.2).

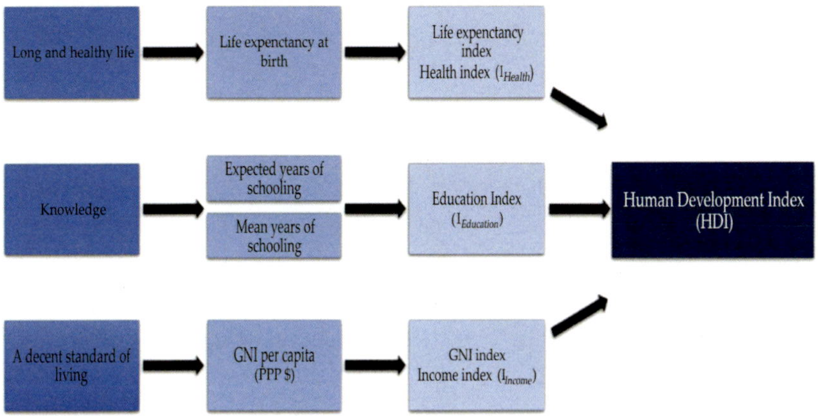

Figure 8.2. The underlying principles behind the Human Development Index

Source: Leon-Castro E, Blanco-Mesa F, Romero-Serrano AM, Velázquez-Cazares M. The Ordered Weighted Average Human Development Index. *Axioms*. 2021; 10(2):87. https://doi.org/10.3390/axioms10020087.

Did you know?

In the latest HDI ranking, from 2022, Switzerland finished first with an HDI value of 0.962. Norway, Iceland, Hong Kong, and Australia rounded out the top five. Meanwhile, the United States was ranked just 21st with an HDI value of 0.921.

8.3.2. Measuring HDI

The Human Development Index (HDI) is a concise assessment of fundamental developmental milestones. An acceptable standard of living, a long and healthy life, and knowledge and understanding are the three characteristics that make up a nation's computed HDI, which is the average of these indexes. First, the three components' geometric mean is determined after each is normalized to a scale between 0 and 1.

- The health aspect of the HDI is measured by life expectancy, calculated at birth in each country, and normalized so that this component is equal to 0 when life expectancy is 20 and equal to 1 when life expectancy is 85.
- Education is measured on two levels: the mean years of schooling for residents of a country and the expected years of schooling that a child has at the average age for starting school. These are each separately normalized so that both 15 mean years of schooling and 18 years of expected schooling equal 1, and a simple mean of the two is calculated.

- The economic metric chosen to represent the standard of living is GNI per capita based on *purchasing power parity* (PPP), a common metric used to reflect average income. The standard of living is normalized so that it is equal to 1 when the GNI per capita is $75,000 and equal to 0 when the GNI per capita is $100.

Keyword

Purchasing power parity (PPP) compares economic growth and standards of living in different countries with a common currency/ basket of goods approach.

The final HDI score for every nation is determined by taking the cube root of the product of the normalized component scores, which yields the geometric mean of the three components (Figure 8.3).

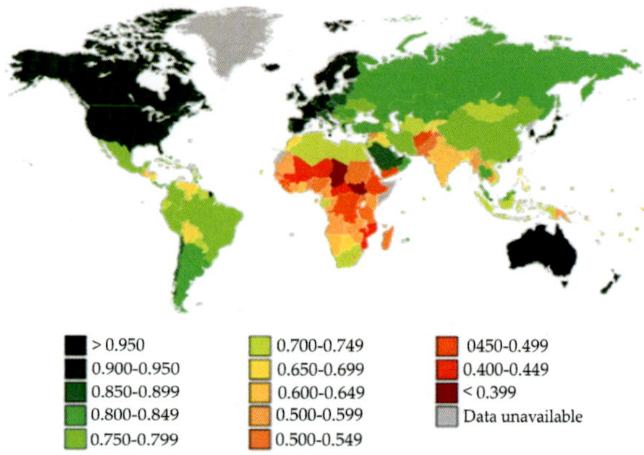

Figure 8.3. World map of countries or territories by HDI scores in increments of 0.050 (based on 2021 data, published in 2022).

Source: By Allice Hunter – United Nations Development Program: Human Development Report 2021–22: Uncertain Times, Unsettled Lives: Shaping our Future in a Transforming World. Human Development Reports 272–275. United Nations Development Program (8 September 2022). Archived from the original on 8 September 2022. Retrieved on 8 September 2022.Blank map: File: Blank world map (Miller cylindrical projection).svg, CC BY-SA 4.0, https://commons.wikimedia.org/w/index.php?curid=122816356.

8.3.3. HDI Rankings

Northern European nations consistently receive the highest HDI scores, with the majority of the continent's lowest-scoring countries being located in Africa. The top 25 countries by the latest HDI rankings (as of 2023) are shown in Table 8.1 and the bottom 5 countries are shown in Table 8.2.

Table 8.1. The Top 25 Countries by the Latest HDI Rankings (as of 2023)

Top 25 HDI Rankings		
Rank	**Country**	**HDI Score**
1	Switzerland	0.962

2	Norway	0.961
3	Iceland	0.959
4	Hong Kong	0.952
5	Australia	0.951
6	Denmark	0.948
7	Sweden	0.947
8	Ireland	0.945
9	Germany	0.942
10	Netherlands	0.941
11	Finland	0.940
12	Singapore	0.939
13	Belgium	0.937
14	New Zealand	0.937
15	Canada	0.936
16	Liechtenstein	0.935
17	Luxembourg	0.930
18	United Kingdom	0.929
19	Japan	0.925
20	South Korea	0.925
21	United States	0.921
22	Israel	0.919
23	Malta	0.918
24	Slovenia	0.918
25	Austria	0.916

Table 8.2. The Bottom 5 Countries by the Latest HDI Rankings (as of 2023)

Bottom 5 HDI Rankings		
Rank	**Country**	**HDI Score**
187	Burundi	0.426
188	Central African Republic	0.404
189	Niger	0.400
190	Chad	0.394
191	South Sudan	0.385

Source: https://www.investopedia.com/terms/h/human-development-index-hdi.asp#:~:text=The%20Human%20Development%20Index%20(HDI)%20measures%20each%20country's%20social%20and,income%20(GNI)%20per%20capita.

8.3.4. Limitations of the HDI

The HDI has been criticized. It is an assessment of human development that is simplified and, to be fair, constrained. Empowerment movements and general feelings of security are examples of quality-of-life factors that are not particularly reflected in the HDI. In light of these realities, the U.S. N. Additional composite indices are offered by the Human Development Report Office (HDRO) to assess other facets of life, such as inequality problems like racial or gender disparities. It is best to examine and evaluate a nation's HDI in tandem with other factors, like the rate of economic growth, the expansion of employment opportunities, and the accomplishment of initiatives aimed at raising the standard of living in the nation as a whole. Because of the strong correlations between the HDI, its components, and more straightforward measures of income per capita, some economists claim the HDI is essentially redundant. In terms of rankings and values, there is a strong correlation between GNI per capita (or even GDP per capita) and the other two components as well as the overall HDI. They argue that, in light of these robust and steady correlations, it would be more straightforward and understandable to compare per-capita GNI between nations rather than investing time and money in gathering data for the supplementary components that add little to nothing to the overall index.

Excluding several additional components that exhibit strong correlations that raise the possibility that they may represent the same underlying phenomenon is a key component of the composite index design. This is done to avoid adding more potential sources of data errors and to stop inefficient double counting. The inclusion of the components in the HDI presents a challenge because it is conceivable that greater average incomes directly correlate with increased spending on formal education as well as improved longevity and health. Furthermore, there can be significant regional differences in the definitions and metrics of life expectancy and years of education.

Question

Who published Human Development Index?

Answer

The Human Development Index is published by the United Nations Development Program (UNDP). It is the most well-known index of human development.

A nation's degree of prosperity is measured by the UN Human Development Index (HDI), which takes into account both economic and non-economic variables. Two examples of non-economic factors are the level of education and life expectancy. Gross national income (GNI) per capita is used to quantify economic factors. As the U.S. National claims that the HDI enhances our comprehension of the relative state of well-being worldwide, economists have criticized the index for being overly simplistic and having problematic methodology.

Did you get it?

1. **What is HDI composed of?**
2. **How is GDI different from HDI?**

8.4. THE DIMENSIONS OF HUMAN DEVELOPMENT

Learning Objectives

- Understand the role of education in human capital and human capability
- Discuss health and human development
- Focus on gender and human development
- Know human development policy perspectives

Studying the various aspects of human development is essential now that you have a better understanding of its definition and assessment. The sections covering the aspects of human development are as follows: gender, health, education, and human development policy perspectives.

8.4.1. Education: Human Capital and Human Capability

Sen compares the capability approach to Gary S. Becker's definition of human capital. He argues that education, in this context, emphasizes the agency of human beings through skills, knowledge, and effort to enhance production possibilities. The capability approach to education focuses on individuals' capacity to lead meaningful lives and make better choices. These approaches are related as they both address the role of humans and the abilities people attain. For example, an increase in human capital occurs when education boosts a person's productivity in producing goods, which can elevate economic value and the individual's income. However, education can also help individuals improve skills such as reading, communication, decision-making, and gaining respect from others, even without directly impacting production. Therefore, the benefits of education go beyond its role as human capital in *commodity* production, highlighting the greater potential of humans.

Three distinct ways have been identified by Amartya Sen to connect the expansion of valuable capabilities and the significance of education. First, education plays a crucial role in society. For instance, literacy promotes discourse and public debate regarding social and political structures. At the community, national, and household levels, education plays a crucial role in enabling us to participate in decision-making processes. Lastly, since these groups couldn't access power centers or influence resource redistribution without education, education plays a distributive and empowering role in helping disadvantaged, marginalized, and excluded groups organize politically. In general, people can use the advantages of education to benefit both themselves and others. They can also contribute to democratic freedoms and the general well-being of society, which means that education has an interpersonal impact. Global proclamations like the Decade of Education for Sustainable Development, Education for All, and the Millennium Development Goals make an effort to consider education from a perspective that goes beyond simple human development.

Keyword

Commodities are basic goods and materials that are widely used and are not meaningfully differentiated from one another.

Education is essential to human development and empowers people. It not only broadens people's perspectives and opens their minds, but it also makes it possible for them to develop other important skills. The first Human Development Index included literacy rates as an education indicator, and subsequent iterations of the index included education indicators based on enrollment rates, reflecting the human development reports' recognition of the critical importance of education.

8.4.2. Health and the Human Development

The common belief is that good health significantly boosts economic growth by enhancing human capital quality. This strategy views health as a means to an end and economic growth as a means to that goal. Conversely, wealth is argued to be an essential component for achieving desired health outcomes. These methods raise the important question of whether economic expansion is required to improve health, though. Amartya Sen illustrates how social services, particularly health care, and basic education can be prioritized in order to improve health (without economic growth) by using the examples of pre-reform China, Sri Lanka, and Kerala (India). According to him, "one of the most important aspects of a person's life is their health, which is also a critically important component of their abilities, which we have reason to value." Apart from its inherent worth, good health plays a crucial role in fostering economic expansion, fostering academic success and cognitive advancement, opening up career paths and earning potential, and promoting dignity, safety, security, and self-determination.

The human development and capability approach is notable for its focus on the process of creating health. The capability approach recognizes that individuals may prioritize health differently and often weigh these priorities against other aspects of their lives. This approach is supported by a significant body of empirical evidence, highlighting the importance of conversion factors in turning health inputs into valuable health outcomes. Education serves as a key example of a conversion factor in health. Numerous studies indicate that individuals with higher education levels tend to have lower mortality and morbidity rates compared to those with lower education levels. Moreover, children of educated mothers generally experience better health outcomes than children of less educated mothers. Another important conversion factor is the individual's ability within their family or community to assess or enhance a specific resource to achieve a health benefit. Factors such as restrictions on women's physical mobility, unequal access to financial resources, and disparities in property rights can all have adverse effects on health. Therefore, conversion factors encompass a range of external conditions.

Keyword

Under nutrition denotes insufficient intake of energy and nutrients to meet an individual's needs to maintain good health.

Everyone has the right to health, which is a fundamental principle of public health. Health disparities are generally defined as variations in the incidence and prevalence of health conditions and health status among groups. Groups that are marginalized due to socioeconomic status, race or ethnicity, gender, sexual orientation, disability, location, or a combination of these factors are primarily affected by health disparities. In addition to experiencing worse health, members of these groups typically have less access to "social determinants" or conditions (e.g., wholesome food, decent shelter, top-notch education, safe communities, lack of racism and other forms of discrimination) that promote a healthy lifestyle. When these crucial conditions are distributed systematically and unfairly, health disparities are known as health inequities. When everyone has the opportunity to "attain their full health potential" and no one is "disadvantaged from achieving this potential because of their social position or other socially determined circumstance," then health equity, according to the WHO, has been achieved.

Therefore, achieving other capabilities relies on good health, which is fundamental. Social justice can be affected by the uneven distribution of health resources. For example, a high number of low birth weight babies is linked to maternal undernutrition, leading to intrauterine growth retardation. This phenomenon results in a high prevalence of adult illnesses and undernourishment in children. The lack of access to healthy food and physical activity for women has negative effects on society. Health-related deprivation can impact various areas such as education, work, and social interactions. People's overall capabilities are significantly reduced when they are unwell or *undernourished*. Lack of health may be the primary cause of interconnected deprivations. According to the WHO Report from 2005, poverty and chronic illnesses are closely connected. Poor health not only indicates illness but also leads to losses in assets, incomes, productivity, and education, perpetuating the cycle of poverty for the impoverished. Therefore, health deficiencies are perpetuated.

8.4.3. Gender and the Human Development

Women often face unequal opportunities compared to men in various societies. There continues to be a prevalent pattern of inequality between men and women in terms of access to healthcare, education, and nutrition, and especially in their participation in political and economic domains.

Amartya Sen has identified seven different types of gender inequalities presently existing in this world.

1. Mortality inequality: In certain parts of the world, the disparity in life expectancy between men and women is extreme, manifested in the disproportionately high death rates of women and the resulting male population relative to the proportion of women in societies where there is little to no discrimination against women when it comes to nutrition and health care. Inequality in mortality has been widely noted in Asia, particularly China and South Asia, as well as North Africa.
2. Natality inequality: In many male-dominated societies, there is a preference for boys over girls. This can lead to gender inequality, where parents wish for a boy instead of a girl when their child is born. Once upon a time, this would have been considered a wish, but sex-selective abortion has become widespread in many nations due to the availability of contemporary medical techniques that allow for the determination of the fetus' gender. It is most common in East Asia, specifically in China and South Korea, but it is also common in Singapore and Taiwan, and it is starting to become recognized as a statistically significant phenomenon in South Asia and India. This sexism is sophisticated. Sen first used the phrase "missing women" to refer to the phenomenon of unequal sex ratios. In its 1974 report "Towards Equality," the Committee on the Status of Women in India (CSWI) considered sex ratios as a composite indicator of women's status in any given society.
3. Basic facility inequality: Women can still be treated unfairly in other contexts, even in cases where demographic traits do not manifest a strong anti-female bias. Although Afghanistan is the only country in the world where the government actively works to prevent girls from attending school (combining this with other aspects of extreme gender inequality), girls have far fewer educational opportunities than boys in many other countries in Asia, Africa, and Latin America. Other inadequacies exist in the fundamental resources accessible to women, ranging from fair participation in community social functions to encouragement to develop one's innate talents.
4. Particular opportunity inequality: Even in cases where there aren't many differences in access to basic services, like education, young women may have significantly fewer opportunities than young men to pursue higher education. In fact, even in some of the wealthiest nations on earth, such as those in Europe and North America, gender bias is evident in professional training and higher education.

5. Professional inequality: Women frequently face greater obstacles than men when it comes to employment and promotions at work. Even though Japan is a fairly egalitarian nation in terms of population, access to basic amenities, and even higher education, women appear to face greater obstacles than men in their pursuit of higher-level jobs and occupations.
6. Ownership inequality: Property ownership can be highly unequal in many societies. Even fundamental possessions like houses and land can be distributed asymmetrically. Women's voices can be diminished by the lack of property rights, which can also make it more difficult for them to participate in and succeed in commercial, economic, and even certain social activities. With some regional variances, this kind of inequality is present throughout most of the world. For instance, in the state of Kerala, matrilineal inheritance has long existed for a powerful segment of the population known as the Nairs, even though traditional property rights have favored men in the majority of India.
7. Household inequality: Gender relations within the family, or household, frequently exhibit fundamental inequalities that can take on a variety of shapes. In situations where there are no overt indications of anti-female bias in areas such as education, preference for sons, survival, or even advancement to higher executive positions, family dynamics can still be highly unequal when it comes to how the burden of childcare and housework is divided. For instance, it's common in many societies to assume that women can work only if they combine work with various unavoidable, unequally shared household duties. Men, on the other hand, are typically expected to work outside the home.

Remember

Matrilineality is the tracing of kinship through the female line. It may also correlate with a social system in which each person is identified with their matriline, their mother's lineage, and which can involve the inheritance of property and titles.

Question

What do you understand by Human Rights?

Answer

Human Rights is a claim to a fundamental benefit that should be enjoyed universally by all people everywhere based on equality and non-discrimination.

There are significant differences in the ways that women share the benefits of human development and the extent to which they contribute to it because of this unequal status. The human development approach contradicts development models that rely on the antiquated trickle-down theory and measure benefits only in terms of money. The trickle-down theory states that the advantages that are distributed to the top of social structures, such as community organizations, eventually trickle down to every member of the community. However, we cannot assume that community development will benefit every member of the community because power and status govern relationships both within and between households. Women and men do not share the same rights in the home, and males and females rarely split household benefits equally.

Keyword

A **non-governmental organization** is an organization that generally is formed independent of the government.

But in many of these nations, women's lower status is associated with their lack of control over assets, which results in systematically reduced access to health and education facilities, community governance, and less than ideal involvement in economic decision-making. Women's low status in society is also influenced by their exclusion from public decision-making, their lack of intra-household bargaining power, and the absence of the kind of kin support that men have in their villages after marriage. Both men and women must inevitably make decisions in the various economic activity domains that involve allocating resources and labor in a household. The ability to negotiate within a family is influenced by a person's ownership and control over assets, availability of work and other sources of income, support from **non**-*governmental organizations* (NGOs), state assistance, social norms, and perceptions of needs, contributions, and other factors determining deservingness. The efforts may also be negatively impacted by the prevalence of patriarchal relationships in the marketplace, which extend to credit availability.

The human development approach is sensitive to aspects of discrimination, such as lack of autonomy in making decisions about their lives and influence over decision-making within the family, community, and country, that are particularly significant in the lives of women but have nothing to do with income and economic growth. The human development approach is also capable of delving into more complicated topics, like the unequal distribution of unpaid labor that limits the options available to women. These problems and their underlying causes undoubtedly need to be addressed given the limitations placed on women's agency in almost all societies by social institutions like the family, political institutions like male-dominated political parties, and social norms like women's caregiving responsibilities. Gender analysis has maintained the approach's vitality, especially in the development of its agency-related aspects.

8.4.4. Human Development: Policy Perspectives

When development priorities and challenges change over time and across national borders, the human development approach is a "robust paradigm" that can be applied. Five components of an all-encompassing agenda for human development are suggested, taking into account the difficulties that the majority of nations are currently facing. These ideas represent what could be referred to as a "New York Consensus" because they are included in numerous UN agreements.

> **Keyword**
>
> **Inclusion** in education refers to all students being able to access and gain equal opportunities to education and learning.

1. Priority to "social development" with the goals of expanding education and health opportunities.
2. Economic growth that generates resources for human development in its many dimensions.
3. Political and social reforms for democratic governance that secure human rights, so that people can live in freedom and dignity, with a greater collective agency, participation, and autonomy.
4. Equity in the above three elements with a concern for all individuals, with special attention to the downtrodden and the poor whose interests are often neglected in public policy, as well as the removal of discrimination against women.
5. Policy and institutional reforms at the global level that create an economic environment more conducive for poor countries to access global markets, technology, and information.

There is a combination of new and old priorities in this five-point agenda. Given that the majority of developing nations have low rates of literacy and that basic survival and health are far from guaranteed, social development is still crucial. Since low growth in developing nations is a significant barrier to human development—more than sixty countries concluded the 1990–2000 decade poorer than when it began—economic growth is another area that continues to receive attention. The human development approach has also undergone a significant evolution at the same time. The HDRs prioritized public spending on health and education in the early 1990s. Priorities in those areas now include resource level as well as service quality, efficiency, and equity of delivery (for which governance reforms are frequently a prerequisite). Higher education standards than those found in elementary school are necessary in today's competitive global markets.

Ensuring the equitable and effective provision of basic services is also greatly aided by institutional reforms that provide the impoverished with the ability to oversee the utilization of local development funds. Most significantly, the HDRs have increased attention to social and political structures that would provide the impoverished and marginalized groups—like women—more influence over public policy and enable them to advocate for their own interests. How international organizations can be redesigned or established to operate on democratic principles, requiring the *inclusion* and participation of all nations and people, is currently another crucial question. For both the instrumental and intrinsic values of education and health, for instance, what the state could do to increase capabilities in these areas constituted a significant pillar of a human development strategy.

The way that the human development approach has evolved over time shows how adaptable it is to new issues and how willing it is to tackle novel policy problems. The human development agenda's policy priorities have changed as a result of these shifting conditions. Public investments have been replaced with incentives, economic measures have been replaced with democratic politics, education and health have been replaced with political and civil liberties, and social and economic policies have been replaced with participatory political institutions. They also acknowledge that, in the current fast-globalizing world, people's capacity for collective action will become more and more crucial in determining how things develop. It is not surprising that state action and changes in planning priorities were at the center of human development advocacy in 1990.

Did you get it?

1. **What are the dimensions of development?**
2. **Explain the significance of human resource development in the context of human development.**

8.5. HUMAN CAPITAL IN ECONOMIC DEVELOPMENT

Learning Objectives

- Explain the role of human capital in economic development
- Know the physical/passive factors of an economy
- Discuss how is human capital formed.
- Focus on investments in human capital
- Understand problems of human capital formation in LDCs

According to contemporary economists, natural resources (i.e. minerals found in forests, the climate, the availability of water, energy sources, etc.) have a significant impact on a nation's economic development. A nation with an abundance of natural resources can develop faster than one with a shortage of those resources. However, a plentiful supply of resources is not a sufficient precondition to account for every facet of economic expansion.

People are the ones who create and run economies. These individuals must be able to carry out the tasks necessary to establish such an economy. Economic growth and decline are contingent upon the size of the population. This is known as human capital, and understanding how populations affect an economy's growth or decline is essential to having a full understanding of the world.

8.5.1. Role of Human Capital in Economic Development

Economic growth is fundamentally derived from human capital. It is a source of both improved technology and higher productivity. The rate at which human capital is developing distinguishes developed nations from developing ones. The developing world needs human capital to staff newly created and growing government positions, implement innovative agricultural and land-use plans, create new communication channels to advance industrialization, and establish educational institutions. Prof. Galbraith is correct when he states that "investments in men and improvements brought about by improved men now account for a larger part of economic growth."

The skills, education, and health gained via on-the-job training and education are referred to as human capital. It is described as "the skill and knowledge of human beings" by Michael Pakistan Park. The "endowment of abilities to produce that exists in each human being" is another definition of it.

How Can a Country Increase Human Capital?

- It can be increased through formal education
- On-the-job training
- Improved health and psychological well-being.

To be more exact, a nation's population is considered to have more human capital if they are skilled, well-nourished, well-educated, and in good health. Investing in human beings is a way for developing nations to improve their health, social skills, programming abilities, and ideals. The goal of these expenditures is to raise productivity. The success of economies rests on the advancement of human capabilities. But human capital doesn't just appear out of nowhere. We need to take into account the physical and passive elements that relate to a nation's capacity to make these *investments* to comprehend this complicated subject better.

Keyword

Investments are also made to benefit from future appreciation in the value of an asset. Investment is a purchase of goods that is future-oriented, aimed at earning income in the future or creating wealth in the future.

8.5.2. Physical/Passive Factors of an Economy

Physical elements are regarded as "passive factors" in economic expansion. They depend on one another and are not independent of one another. These people are thought to be "active factors" in economic growth. The quantity of land that is available in each of those areas is a passive factor, but the active factors of a nation include crucial metrics like the size of the population's growth rate in both urban and rural areas. A country's cultural and economic advancement is greatly influenced by the quality of its population, as evidenced by health, education, and technology levels; however, the resources needed to attempt these ambitious improvements—namely, money and land—are unavoidably linked to these factors.

A nation that has invested in its people's skills and knowledge can take advantage of its natural resources, create political, social, and economic institutions, and advance its own development. Having said that, a nation will find it difficult to achieve the fast increase in human capital that it hopes for if it ignores the passive elements that shape these objectives.

8.5.3. How Is Human Capital Formed?

The process of improving the labor force's productivity through increased education, skill development, improved health, and higher levels of notarization among employed people is known as "human capital formation."

The five ways of developing human capital:

1. Provision of health facilities that affect the life expectancy, strength, vigor, and vitality of the people
2. Provision of on-the-job training, which enhances the skill of the labor force
3. Arranging education at the primary, secondary, and higher levels
4. Study and extension programs for adults
5. Provision of adequate migration facilities for families to adjust to changing job opportunities

Remember

Although investment in human capital tends to produce more growth, it doesn't necessarily mean jobs are available for the newly educated workers. Also, geography plays a role when it comes to job openings and the movement of labor. If job openings are located in the northern part of a country, but the skilled labor is in the south, growth could be hindered due to the cost of moving or the lack of desire to move.

8.5.4. Investments in Human Capital

Economic growth and human capital are positively correlated because investments often increase productivity. Training a workforce is a form of investment, but the investment is in human capital rather than capital goods like machinery.

8.5.4.1. Government Investment

Increasing a nation's population's skill and educational levels is largely dependent on the role played by its governments. Certain governments take an active role in enhancing human capital by providing free higher education to citizens. These governments understand the importance of education in fostering economic development and growth. Higher incomes for workers with greater education or skill levels typically lead to higher consumer spending, which boosts economic growth.

8.5.4.2. Corporate Investment

To increase revenue and productivity, businesses also invest in human capital. Let's take an example where a worker at a technology company attends internal seminars and on-site training to become a computer programmer. A portion of the cost of higher education tuition is covered by the employer. After completing the training, if the employee stays on staff, they might come up with fresh concepts and new products for the business. Later in their career, the employee may decide to leave the company and use the skills they acquired to launch a new business.

Economic growth is the ultimate result of the initial investment in human capital, regardless of whether the employee starts a new business or stays with the company.

8.5.5. Problems of Human Capital Formation in LDCs

Although investing in the development of human capital has numerous advantages for LDCs (less developed countries), it is not a simple task. Big populations have big problems to solve.

1. **Faster increase in population**: Nearly every developing nation on the planet, including Pakistan, is experiencing a faster rate of population growth than the rate at which human capital is being accumulated. Consequently, these nations are not effectively utilizing the resources allocated to education, which has accounted for 2.5% of LDC GDP over the past five years.
2. **Defective patterns of investment in education:** Governments in developing nations are prioritizing primary education to enhance literacy rates. Secondary education, despite its importance in imparting essential skills for economic growth, remains undervalued. The proliferation of universities in both public and private sectors poses another challenge in education investment. The establishment of universities in these nations comes at a significant cost. Widespread failures in elementary, secondary, and higher education also result in the misallocation of limited resources needed for other forms of development.
3. **More stress on the provision of buildings and equipment:** The fact that politicians and administrators place more emphasis on building infrastructure and providing equipment than on hiring skilled labor is another significant issue that developing nations face when investing in human capital. It has been noted that foreign-trained physicians and educators are frequently assigned to remote locations where their skills are underutilized. This misallocation of funding for education may have a detrimental impact on economic expansion.
4. **Shortage of health and nutrition facilities**: There is a scarcity of medical supplies, medications, certified physicians, trained nurses, and other medical professionals in less developed nations. Millions of people are at risk if there are fewer healthcare facilities available. The population has to deal with poor living conditions, contaminated water supplies, high rates of births and deaths, urban shantytowns, illiteracy, etc. People's lives are shortened and their health is impacted by all of these deficiencies. As a result, human capital grows less rapidly.
5. **No facilities for on-the-job training**: To develop new or better skills, on-the-job training, also known as in-service training, is crucial. The end effect is an increase in human capital due to worker productivity and employee knowledge. The most crucial factor in the effective use of human resources is employee competence.

6. **Study programs for adults:** Adult education programs can also be implemented to increase a nation's literacy rate. Adult education initiatives have been implemented in some developing nations, including Pakistan. They provide basic education, which helps small industrialists and farmers become more skilled. Unfortunately, this initiative was a complete failure because adults showed no interest in receiving this type of instruction.
7. **Halfhearted measures for the promotion of employment**: The percentage of people who are unemployed or underemployed is extremely high in most parts of the world. Adequate investment in human capital is essential to enhance employment and reduce underemployment. This is notably lacking in LDCs. One positive example is the various measures that the Pakistani government has taken to improve job prospects in the country, such as establishing the SME Bank to promote self-employment at the local level. This encourages both foreign and domestic investment, leading to an increase in job opportunities. Furthermore, more technical and vocational training facilities can be established as a result.
8. **Failure to plan for the optimal use of manpower**: Less developed nations have limited manpower planning due to the scarcity of reliable data. Consequently, there is a mismatch between the supply and demand for specific skills. Many competent and highly qualified workers remain underemployed as a result. "Brain drain" occurs when underemployed and unemployed graduates and postgraduates become frustrated and unhappy, prompting talented individuals to leave the country in search of better opportunities abroad. These developing nations have experienced a significant loss of human resources.
9. **Neglect of agriculture education:** Teaching farmers how to use modern agricultural practices receives very little attention in least developed countries (LDCs), where agriculture is the primary economic sector. Farmers cannot increase agricultural output and maintain supply and demand equilibrium unless they receive agricultural education and training.

Question

Why does human capital matter to an economy?

Answer

The knowledge, skills, and creativity of a company's human capital is a key driver of productivity. Developing human capital allows an economy to increase production and spur growth.

Did you get it?

1. **What is the role of human capital in economic development?**
2. **How is human capital formed?**
3. **What Are the Economic Benefits of Human Capital?**

ROLE MODEL

MAHBUBUL-HAQ

Pioneering a Development Philosophy for People

Mahbubul-Haq (24 February 1934– 16 July 1998) was a Pakistani economist, international development theorist, and politician who served as the Minister of Finance of Pakistan from 10 April 1985 to 28 January 1986, and again from June to December 1988 as a caretaker. Regarded as one of the greatest economists of his time, Haq devised the Human Development Index, widely used to gauge the development of nations.

After graduating with a degree in economics from the Government College University in Lahore, he won a scholarship to the University of Cambridge in England, where he obtained a second higher degree in the same field. He later received his Ph.D. from Yale University in the United States and subsequently conducted postdoctoral research at the Harvard Kennedy School. Haq returned to Pakistan to serve as the chief economist of the Planning Commission throughout the 1960s. In 1970, after the fall of Ayub Khan, Haq moved to Washington, D.C. to serve at the World Bank as Director of Policy Planning until 1982, where he played a major role in reorienting its approach to assisting development in low-income countries.

He returned to Pakistan in 1982, and in 1985 assumed the position of Finance Minister with the Government of Pakistan, and oversaw a period of economic liberalization in the country. In 1989, he moved back to the United States, where he served as the special adviser to the United Nations Development Program (UNDP) under its head, William Henry Draper III. At the UNDP, Haq led the establishment of the Human Development Report and the widely-respected Human Development Index (HDI), which measures development by well-being, rather than by financial income alone. He returned to Pakistan in 1996 to establish the Human Development Centre in the capital city of Islamabad.

Haq is considered to have had a profound effect on global development. His 1995 book, *Reflections on Human Development*, is said to have opened new avenues to policy proposals for human development paradigms, such as the United Nations Global Compact that was formed in 2000. Amartya Sen and Tam Dalyell judged Haq's work to have "brought about a major change in the understanding and statistical accounting of the process of development." *The Economist* called him "one of the visionaries of international development." He was widely regarded as "the most articulate and persuasive spokesman for the developing world."

Early Life and Education

Haq was born into a Punjabi Muslim family on 24 February 1934 in the city of Gurdaspur, Punjab Province, British India (now located in the Republic of India). His teenage years saw widespread intercommunal violence and forced migration following the independence of India and Pakistan from British rule in August 1947. He and his family migrated from India to the newly-created state of Pakistan following the partition of India; Haq stated that they narrowly escaped being killed in one of the refugee trains heading to Pakistan. After reaching Lahore, Haq was given government-sponsored housing and decided to continue his education.

In 1953, he graduated with a degree in economics from Lahore's Government College University. He later earned a scholarship to attend Cambridge University, where he earned another BA in the same discipline alongside Indian economist Amartya Sen, with whom he formed a close friendship. After renewing his scholarship, Haq went to the United States for his doctoral studies at Yale University and obtained a PhD. Later, Haq carried out postdoctoral work at Harvard University in 1960–61.

Career

An early proponent of economic liberalization who, in later years, argued that poor countries failed to prosper because they neglected the basic development of their people — *New York Times*. Upon returning to Pakistan in 1957 at the age of 23, Haq joined the Planning Commission as Assistant Chief while it prepared its first Five-Year Plan. Influenced by the dominant economic thought in American academia, Haq advocated capitalism as the economic base of the national economy and helped guide the government to apply free-market principles to boost the economy. This approach was wholeheartedly embraced by the military government of General Ayub Khan after it came to power in October 1958. By the 1960s as Chief Economist of the Planning Commission, Haq was delivering speeches all over the country in support of these economic policies. While the international community was applauding Pakistan as a model of development, Haq developed concerns that all was not well with the distribution of the benefits of growth. Rapid economic development made Haq's team doubt the long-term viability of such a pattern of growth, and he increasingly supported heavier taxation of the asset-owning classes. In a widely reported speech to the Applied Economics Research Centre at the University of Karachi in April 1968, Haq alleged that "22 industrial family groups had come to dominate the economic and financial life-cycle of Pakistan and that they controlled about two-thirds of industrial assets, 80% of banking, and 79% of insurance assets in the industrial domain." The list included the Dawood family of Dawood Group, Saigols of Saigol Group, Adamjees of Adamjee Group, Colony, Fancy, Valika, Jalil, Bawany, Crescent, Wazir Ali, Gandhara, Ispahani, House of Habib, Khyber, Nishat Group, Beco, Gul Ahmed Group, Arag, Hafiz, Karim, Milwala, and Dada.

These revelations played a major role in mobilizing millions in a massive grassroots protest movement that led to Field Marshal Ayub Khan's overthrow in March 1969. Following Ayub's fall, Haq accepted an invitation from Robert McNamara, president of the World Bank, to serve as his Director of Policy Planning. During his tenure (1970–82), Haq influenced the Bank's development philosophy and lending policies, steering more attention toward poverty alleviation programs and increased allocations for small farm production, nutrition, education, water supply, and other social sectors. He wrote a study that served as a precursor to the basic needs and human development approaches of the 1980s.

While working at the World Bank, Haq was invited by Prime Minister Zulfiqar Ali Bhutto to join the Ministry of Finance but ultimately refused as he had strong opposing views on Bhutto's program of nationalization. In 1973, Bhutto again asked Mahbub to return to Pakistan and join his administration in devising a strategy that would lift a large number of Pakistanis out of poverty and stagflation, but their major differences persuaded Haq not to return.

In 1982, Haq returned at the request of General Zia-ul-Haq's military government, where he assumed directorship of the Planning Commission. In 1983, Haq was appointed Minister of Planning and Development. According to Parvez Hasan, 'under Mahbub's direction, the Planning Commission became once again a lively place and began to exert powerful influence on social sector issues, including education and family planning, much neglected in earlier Zia years.'

In 1985, President Zia oversaw a partial return to democracy with so-called 'non-party' general elections, and Haq was sworn in as Minister of Finance, Planning, and Economic Affairs in the PML government of Mohammed Khan Junejo. Haq is credited with significant tax reforms, deregulation of the economy, increased emphasis on human development, and several initiatives for poverty alleviation. Despite this major acceleration in social spending, Haq was forced to resign in January 1986 due to protests regarding his reforms. He was reappointed as Finance Minister in the caretaker administration established by General Zia-ul-Haq after he dismissed the Junejo government in May 1988. Haq's term ended when the PPP government of Benazir Bhutto was sworn in following the general elections of November 1988.

In 1989, he was appointed as Special Advisor to the UNDP Administrator William Draper in New York City to produce the first Human Development Report. In this capacity, Haq initiated the concept of Human Development and the Human Development Report as its Project Director. He led a team of international scholars including Amartya Sen, Paul Streeten, Inge Kaul, Frances Stewart, and Richard Jolly to prepare annual Human Development Reports.

In 1996, Haq founded the Human Development Center in Islamabad, Pakistan — a policy research institute committed to organizing professional research, policy studies, and seminars in the area of human development, with a special focus on South Asia. In acknowledgment of his contributions, the Human Development Centre, Islamabad was officially renamed following his death as the Mahbubul Haq Human Development Centre on 13 December 1998, with Mrs. Khadija Haq as president.

Death

Haq died on 16 July 1998 in New York City at the age of 64, leaving behind his wife Khadija Haq, son Farhan and daughter Toneema.

CASE STUDY

THE IMAGINARY COUNTRY OF TARTAL

Tartal is a developing nation with a high standard of living. It is among the most populous nations in the world and the eighth largest by land area. With a per capita GDP of $19,000 (PPP US$), Tartal is not considered an "advanced economy" by the IMF; however, its 2% annual growth rate indicates a promising path toward development in the future. However, a large number of cultural and ethnic groups have not equally shared in the growth performance, as evidenced by the statistics.

In 1887, Tartal achieved independence, and in 1901, a new constitution was ratified. Following a turbulent political history that spanned the majority of the 20th century, Tartal overthrew a 20-year dictatorship in 1982 and instituted democracy in 1985. Ever since the nation has endeavored to utilize its copious amounts of land rich in resources to meet the lofty goals of economic development and modernization in the context of globalization.

With 150 million people living there, 60 million of them are under 25, and the population is growing at a rate of 1% per year. Although a recent migration to large cities has resulted in a dramatic increase in the urban population, nearly 70% of its inhabitants live in rural areas. With a population of 10 million, the capital of Mani is the largest metropolitan area; Safa, with 7 million people, is the second largest. The fact that both cities are located on the Pacific coast highlights the contrast between the affluent, urban coast and the rural, underdeveloped interior. About half of the population is classified as indigenous, and the other half as "white" or "mixed," deriving from European ancestry. Tartal has one of the greatest economies in the world, but it also has high rates of illiteracy, poverty, and chronic malnutrition. Even though the average income in the nation is very high, it is very unevenly distributed among the population.

Tartal's income distribution is unequal, with the richest 10% of the population receiving 30% of the nation's income while the poorest 10% of the population only receives 1% of the nation's income. Furthermore, the "freedoms" that women in this nation enjoy in the social, cultural, and economic spheres are severely limited. Women hold just 8% of the parliamentary seats.

Environmental concerns have suffered as a result of the rapid economic expansion. Industrial effluents pollute the environment, excessive chemical use in agriculture affects agriculture, and deforestation occurs. The nation's rich natural resources are being neglected, posing a threat to the country's biodiversity. The democratic system of government in the nation is not entirely participatory; people lack a voice in important decisions that impact their daily lives and the media is suppressed. Although the adult literacy rate is estimated to be 65%, only 30% of indigenous people are literate. The critical lack of indigenous participation in government is further compounded by this illiteracy.

FOCUS ON CAREERS

ECONOMIST

Economists forecast future societal and financial events by examining a wide range of economic issues and historical trends. If you're interested in a career in economics, there are some basic steps you can take to start building foundational competencies and making a good living. You can assess if you have the required training and experience by learning how to apply for this position.

An economist studies the reasons behind human decision-making and is interested in using statistics to boost profits, enhance public policy, or conduct research. Professionals who study development, acquisition, and wealth transfer are known as economists, and their goal is to apply their knowledge to manage the dynamics of modern society. In India, an economics career entails overseeing the physical progress of nations, government-implemented social reforms, and the ascent and decline of authoritarian regimes.

Understanding and utilizing the power of economics are important tasks for economists. In India, an economics career entails data collection and analysis, trend research, and the assessment of economic issues pertaining to resources, goods, and services. If someone chooses to become an economist, their job will involve applying both qualitative and quantitative economic analysis to various industry niches. For example, the environment, health, development, and education.

What does an economist do?

Three definitions of economics are provided by the American Economic Association (AEA): the study of scarcity, the study of how people use resources, and the study of decision-making. Most people associate economics with Wall Street bankers, but this is a narrow understanding of economists.

In order to make sense of historical trends, separate news from noise in today's headlines, and make predictions about the future, economists work in a variety of fields of study, including health care, banking, employment rates, law, business trends, finance, wealth management, politics, taxation, criminal justice, and media outlets of all stripes.

Economists conduct research and study business cycles, labor statistics, market trends, and other economic issues. They examine economic trends in their particular industry using both qualitative and quantitative economic analysis. A lot of economists dedicate their professional lives to studying trends or doing data analysis. They might look at employment figures or project market movements.

All of these skills are expertly combined by economists to assess the financial performance of different businesses or sectors. More in-depth knowledge of economic matters may even open doors for economists looking to contribute to scholarly journals.

Some economists who specialize in business economics or economic issues write about economic issues and offer suggestions for solutions.

Qualifications to Become an Economist

It is advised that students who envision a career in economics pursue an economics degree. An advanced degree, like a Ph.D., is regarded as a desirable credential by companies in the field of economics. Many of the aforementioned specialties have graduate programs offered by a large percentage of colleges and universities, giving students the option to choose the program that best suits them. It is recommended that undergraduates studying economics take a variety of related courses in fields like computer science, statistics, mathematics, and survey design.

Graduate students can obtain experience through various internships, and one requirement for those looking to work for the federal government is to have completed at least three hours of advanced math in addition to the required twenty-one hours of general economics. Most public and private companies require a bachelor's degree in economics for entry-level positions. Most high-level jobs naturally require a Ph.D., while economists hoping to become college professors will need a master's degree.

Job Outlook for Economists

The majority of economists will work in entry-level positions at organizations like research firms, financial institutions, and government agencies, all of which can lead to other career opportunities. It is anticipated that employment for economists will increase at the natural average rate in the future, with the private sector accounting for the majority of the growth. More specifically, consulting firms and scientific research companies are believed to have the highest demand for economists as the need for specialization in a variety of economic fields continues to rise.

A CLOSER LOOK

FINANCIAL STABILITY

The definition of "a condition in which the financial system is not unstable" is financial stability. It may also refer to a situation in which the financial system's three constituent parts—financial markets, financial infrastructure, and financial institutions—are all stable.

- Stability of financial institutions' refers to a condition in which individual financial institutions are sound enough to carry out their financial intermediation function adequately, without assistance from external institutions including the government.
- Stability of financial markets' means a condition in which there is no major disruption of market transactions, with no significant deviation of financial asset prices from economic fundamentals, thereby enabling economic agents to raise and operate funds with confidence.
- Stability of financial infrastructure' refers to a condition in which the financial system is well structured to ensure the smooth operation of market discipline, and that both the financial safety net and the payment and settlement system are running effectively.

Financial stability can be more broadly defined as "a condition in which the financial system can facilitate real economic activities smoothly and is capable of unraveling financial imbalances arising from shocks."

In addition to being necessary for price stability, which is the central bank's policy objective, financial stability is also necessary for the economy to grow healthily. This is because financial instability comes with high costs for an economy since it can lead to financial institutions or corporations going bankrupt and increase the volatility of price variables in the financial markets. Furthermore, because economic agents find it difficult to make logical decisions and resource allocation becomes less efficient, economic development may be constrained during this period. The progress of financial liberalization has allowed many countries to reap the benefits of fast-growing financial industries since the 1980s. On the other hand, because of significant costs associated with financial instability or crises, they have also gone through times of sharply slower economic growth.

In light of this, many nations have begun to prioritize financial stability when putting their policies into action. Financial stability is receiving more attention as a result of the recent emergence of new factors that have the potential to cause it, such as the strengthening of international financial sector ties and the rapid development of sophisticated financial instruments.

SUMMARY

- Development and economic growth are crucial for the long-term sustainability of a nation's economy. Economic growth is generally seen as a good thing as it shows that the economy is generating more goods and services, leading to increased output.
- A nation's economy is usually considered to be doing well when it is experiencing economic growth. However, this growth does not always result in an enhancement of the nation's overall standard of living or economic prosperity. This is where economic development can be a valuable metric.
- "Traditional society" is the name of the initial phase. Rostow coined the term to describe societies with limited technological progress.
- Economic growth is often assessed using GDP and GNI per capita as they provide a more precise indication of a country's income compared to combined GDP and GNI figures.
- Growth and development, unfortunately, are not always straightforward processes. An economy might face challenges that hinder its sustainable expansion and development.
- To foster economic growth and development, different macroeconomic and microeconomic policies can be employed. Supply-side policies, monetary policies, and fiscal policies are some examples of potential measures.
- The approach to human development is diverse in nature. Making it possible for people to take an active role in their own lives is the main objective of human development.
- The ideal of equality, according to which every person should have an equal voice in civic and political life as well as equal rights and entitlements to human, social, economic, and cultural development, is encompassed by the principle of equity.
- Efficiency is the least expensive way to achieve objectives by making the best possible use of institutional, material, and human resources to maximize opportunities for both individuals and communities and boost productivity.
- Engagement and participation in social and political life is a crucial component of human development.
- Human development challenges the notion that economic growth can be sustained over the long run and works to make sure that resources are used in a way that both satisfies current needs and protects the environment so that needs of future generations can be satisfied as well.
- Human development is categorized into three stages: higher, medium, and lower-level development groups are established for each country. The Human Development Index (HDI) gives an overview of average performance in three key areas: living a long and healthy life, having knowledge, and enjoying a reasonable standard of living.

- The Human Development Index (HDI) was designed to prioritize individuals—or, more accurately, their potential to achieve fulfilling careers and personal lives.
- The Human Development Index (HDI) is a concise assessment of fundamental developmental milestones.
- Consistently, Northern European nations have the highest HDI scores, while most of Africa's countries rank lowest.
- The HDI, criticized for its simplified and constrained assessment of human development. Empowerment movements and general feelings of security are examples of quality-of-life factors that are not particularly reflected in the HDI.
- The capability approach to education centers on people's capacity to live meaningful lives and make better meaningful choices in their lives.
- The conventional wisdom contends that health has a substantial and positive impact on economic growth because it improves the quality of human capital. This strategy views health as a means to an end and economic growth as a means to that goal.
- Women never have the same opportunities as men in any society. There is still a pervasive pattern of inequality between men and women when it comes to their access to health care, education, and nutrition, and even more so when it comes to their involvement in the political and economic spheres.

MULTIPLE CHOICE QUESTIONS

1. Which of the following is not one of the four pillars of Human Development?

a. Equity

b. Sustainability

c. Accessibility

d. Productivity

2. Development of a few regions by individuals brought about in a short span of time leads to:

a. Malnutrition

b. Poverty

c. Ecological degradation

d. All of these

3. Which of the following steps can be taken to improve the level of human development?

a. Increasing literacy levels

b. Providing health facilities

c. Having more enrollment ratio

d. All of these

4. Which of the following indicators is not used to measure Human Development in HDI?

a. Purchasing Power

b. Gross Enrolment

c. Vaccination

d. Life Expectancy

5. Which pillar of Human Development stresses on opportunities for future generations?

a. Equity

b. Empowerment

c. Sustainability

d. Productivity

6. Which of the following pairs is not correctly matched?

a. Equity – It refers to making equal access to opportunities available to everybody.

b. Sustainability – It means continuity in the availability of opportunities.

c. Accessibility – It means human labor productivity or productivity in terms of human work.

d. Empowerment – It means to have the power to make choices.

7. According to the Human Development Report, 2018, United Nations Development Program, which two countries have a higher life expectancy at birth than India?

a. Pakistan and Nepal.

b. Pakistan and Bangladesh.

c. Nepal and Bangladesh.

d. Sri Lanka and Myanmar.

8. If women are engaged in paid work, what difference does it make?

a. No difference.

b. No dignity.

c. Their dignity in the household and society increases.

d. Their dignity in the household and society decreases.

9. The Human Development Index compares countries based on which of the following levels of people?

a. Health status.

b. Per Capita Income.

c. Educational level.

d. All of the options are correct.

REVIEW QUESTIONS

1. Give an example of human development?
2. What are the pillars of human development?
3. What are the stages of human development?
4. What is economic development?
5. What is the human development approach to economic development?
6. What does income help you to achieve things you consider as valuable? What do you see as valuable things to do or be in life that income does not influence?
7. If you were to assess the well-being of people in your country, what information would you include? Why?
8. What are the indicators used in the Human Development Index (HDI)?
9. Which countries have the highest HDI?
10. Gender development is an important component of Human Development. Narrate.

Answer to Multiple Choice Questions

1. (c)	2. (d)	3. (d)	4. (c)	5. (c)
6. (c)	7. (c)	8. (c)	9. (a)	

REFERENCES

1. Deneulin, S., & Lila, S., (2009). *An Introduction to the Human Development and Capability Approach: Freedom and Agency.* Human Development Capability Association, United Kingdom: Earthscan.
2. Martha, C. N., (2000). *Women and Human Development: The Capabilities Approach.* Cambridge, UK: Cambridge University Press.
3. Parr, S. F., (2003). The human development paradigm: Operationalizing Sen's ideas on capabilities. *Feminist Economics, 9*(2, 3), 301–317.
4. Parr, S. F., & Shiva, K. A. K., (2003). *Readings in Human Development.* New York: Oxford University Press.
5. Sen, A., (2001). *Many Faces of Gender Inequality* (Vol. 18, No. 22). Frontline.
6. Sen, A., (2002). Why health equity? *Health Economics, 11*, 659–666.

Index

E

F

G